From DevOps to SecDevOps

Automating Security Across the SDLC

By Dr. Edward K. S. Buckman, PhD, MBCI, CBCP, ITIL

Published by **Nexora Press**

ISBN: 979-8-9991843-2-0

Dedication

To the tireless security engineers, developers, and compliance leaders who believe that speed and safety can, and must, coexist.

Publishing Information

Nexora Press
Johnson City, Tennessee, USA
www.nexorapress.com

Legal Disclaimer

This book is intended for educational and professional development purposes. While every effort has been made to ensure the accuracy and relevance of the content, the author and publisher assume no responsibility for errors, omissions, or any outcomes related to the implementation of concepts discussed herein. Compliance requirements and tools may vary based on jurisdiction, industry, and evolving standards.

Copyright Disclaimer

About the Author

Dr. Edward K. S. Buckman is an expert in cybersecurity, cyber insurance, and risk management with extensive experience in secure software development, business continuity, and IT governance. Holding a PhD in Technology and Innovation Management with a focus on cybersecurity. Dr. Buckman has worked across diverse sectors to design resilient and secure digital infrastructures. He is the founder of Nexora Press, committed to delivering high-quality educational resources in emerging technology domains.

Acknowledgments

Special thanks to the dedicated teams, reviewers, and security professionals whose real-world experience, feedback, and resilience shaped the insights in this book. Your contributions help the industry move forward with clarity and confidence.

Author's Note

Throughout my career in cybersecurity, DevOps, and resilience engineering, I've seen how easily "security" can be treated as an afterthought, something owned by another team, added as an afterthought, or overlooked for the sake of delivery speed.

SecDevOps is the remedy to that mindset. It empowers builders to be protectors.

This book reflects everything I believe about professional software engineering:

- That security should be invisible yet pervasive

- That we must collaborate across silos

- That we can build systems that are both fast and safe

Thank you for reading this book. I hope it becomes part of your toolbox, your team's culture, and your organization's success.

Keep learning. Keep leading. Keep securing.

Preface

In a world where software powers nearly everything, from healthcare systems to financial markets, ensuring that software is secure is no longer optional.

This book was born from a simple, urgent question: What if we could build security into software, without slowing it down?

The answer is *SecDevOps*.

I wrote this book not only for security experts, but also for developers, DevOps engineers, and technical leaders who recognize that software delivery must be both fast and secure, as well as auditable and resilient. This book provides a step-by-step roadmap for integrating security into the core of your software pipelines.

You'll find:

- Real-world tools and walkthroughs
- Hands-on labs and CI/CD integrations
- Maturity models and team strategies
- A security culture built into the code

My goal is to help you become not just a secure coder, but a secure systems thinker.

Let's build the future — one secure commit at a time.

— Dr. Edward K. S. Buckman

Table of Contents

Part I: Foundations of SecDevOps

This section introduces the core principles, lifecycle integration, and essential secure development practices that form the bedrock of SecDevOps.

- Chapter 1: What Is SecDevOps?
  Introduces the concept, history, and value proposition of SecDevOps as a modern evolution of DevSecOps.
- Chapter 2: Secure Development Lifecycle (SDLC)
  Explores how security can be embedded into every phase of the software development process—from planning to deployment.
- Chapter 3: Securing Code Repositories and Git Workflows
  Focuses on access control, commit hygiene, branching strategies, and protecting source code integrity.

Chapter 1: What Is SecDevOps?

From DevOps velocity to secure-by-default development

1.1 Introduction

In today's fast-paced software delivery landscape, speed is no longer enough; security must be built into the process, not bolted on afterward. This is the essence of SecDevOps: a mindset, practice, and culture that embeds security into every stage of the software development lifecycle (SDLC), from planning and coding to deployment and monitoring.

As attacks become more sophisticated and systems become increasingly complex, the industry is transitioning from siloed security to continuous, automated, and integrated security. That shift is called SecDevOps.

1.2 SecDevOps vs. DevSecOps — Is There a Difference?

While many use SecDevOps and DevSecOps interchangeably, some differentiate the two:

Term	Emphasis	Flow	Culture Focus
DevSecOps	Adding security to DevOps	Dev → Sec → Ops	Reactive integration
SecDevOps	Embedding security first	Sec → Dev → Ops	Proactive, security-first mindset

SecDevOps is not about adding security as a barrier; it's about weaving it in as a default expectation. It prioritizes early threat modeling, secure code reviews, and automated scans in the same way DevOps prioritizes automation and continuous integration/continuous delivery (CI/CD).

1.3 The Building Blocks of SecDevOps

To successfully embed security into the DevOps lifecycle, organizations must move beyond simply adopting tools; they must also integrate security into their processes and workflows. True SecDevOps requires a shift in mindset, process, and infrastructure. This shift is supported by a foundation of interconnected pillars that ensure security is not only integrated but also orchestrated across the Software Development Life Cycle (SDLC).

Below are the core building blocks of a resilient SecDevOps strategy. Together, they form a framework for continuous, automated, and team-aligned security implementation:

Component	Role in SecDevOps
Culture	Shared security ownership across teams
Automation	Continuous integration of scans and checks

Component	Role in SecDevOps
Collaboration	Dev, Ops, and Security teams working as one
Monitoring	Observability for security events
Compliance as Code	Audit-ready infrastructure and workflows

1.4 Why SecDevOps Is Critical Today

As software delivery accelerates, so too do the risks associated with insecure development and deployment. Traditional security models, built for slower, monolithic systems, cannot keep pace with today's fast, iterative DevOps pipelines. Attackers exploit this velocity gap, while organizations struggle to secure highly dynamic, cloud-native architectures.

SecDevOps becomes critical because it embeds security directly into the delivery process itself, at speed and scale, making secure software delivery both achievable and sustainable. Below is a range of modern realities that are driving the urgent need for SecDevOps today:

1. **Modern attack surfaces**: Cloud-native, containers, APIs, mobile, and IoT

2. **Compliance and privacy laws**: GDPR, SOC 2, HIPAA, PCI-DSS

3. **Breach response costs**: Prevention is far cheaper than reaction

4. **CI/CD velocity**: You deploy code daily; security must keep up

Fact: According to IBM's 2023 Cost of a Data Breach Report, the average breach cost exceeded **$ 4.45 M.** Automating security in DevOps pipelines reduced breach impact and time-to-detect by over 25%.

1.5 Traditional DevOps Without Security

While DevOps has dramatically improved development speed, automation, and operational efficiency, security has often struggled to keep up. In many organizations, DevOps pipelines are designed to maximize delivery speed, with security usually being left as a separate concern, handled post-deployment or, worse, after an incident has occurred. Without integrated security, DevOps pipelines unintentionally become highly efficient delivery systems for untested, unverified, and potentially vulnerable code. The speed that makes DevOps attractive can also accelerate the introduction of security flaws into production environments. When security is not embedded into DevOps workflows, several predictable and dangerous patterns emerge. Here's what often happens when security is left out of DevOps:

- Vulnerabilities are discovered after deployment

- Developers receive delayed security feedback

- Security reviews block releases last-minute

- Compliance is done manually and inconsistently

SecDevOps solves this by fundamentally redesigning how security fits into modern development. Instead of trying to add security as an afterthought or as a blocker at the end of the pipeline, SecDevOps embeds security directly into the daily workflows, automation, and culture of software teams. It transforms security into a continuous, automated, and collaborative practice that scales with delivery speed.

SecDevOps solves this by:

- Integrating SAST/DAST/SCA tools into CI/CD

- Creating pre-approved, secure templates

- Enforcing policy as code automatically

- Using threat modeling during design

1.6 Common Myths About SecDevOps

As organizations begin to adopt SecDevOps, they often encounter misconceptions that hinder their full implementation. Many of these myths stem from outdated security models, a fear of change, or a misunderstanding of the true nature of integrated security. Clearing up these myths is essential for building the right mindset and getting leadership, developers, and security teams aligned on what SecDevOps is and what it is not.

Myth	Reality
"It slows us down"	Done right, it **accelerates releases** by catching issues early

Myth	Reality
"Security is someone else's job"	Everyone shares responsibility in SecDevOps
"We'll add it later"	Later = breach risk + rework + cost

1.7 Real-World Adoption Examples

While SecDevOps may sound like a modern framework, many leading organizations across industries have already proven its value through real-world adoption. These companies demonstrate how integrating security into DevOps pipelines can enhance delivery speed, mitigate breach risks, and simplify compliance, all while fostering stronger customer trust. The following examples highlight how different organizations have successfully operationalized SecDevOps principles in their unique environments:

- **Netflix** uses "Security Monkey" in CI/CD to detect insecure configurations

- **Google** enforces security linters and secret scanners in developer IDEs

- **Capital One** built DevOps security maturity models and secure baselines

1.8 Chapter Summary

SecDevOps is not simply an evolution of DevOps; it's a necessary transformation that addresses the security realities of today's software-driven world. As this chapter has demonstrated, the rise of cloud-native architectures, the increasing attack surface, and the demand for continuous delivery have made integrated security an indispensable requirement. The key concepts from this chapter provide the foundation for understanding why SecDevOps exists, what challenges it solves, and how organizations can begin adopting its core principles.

- SecDevOps = Security by Design, not by delay
- It's a cultural shift, supported by automation, testing, and policy
- Prevents breaches by bringing security into the SDLC loop
- Speeds up delivery while increasing trust, compliance, and resilience

You don't need to trade speed for security. With SecDevOps, you get both by design.

Chapter 2: Secure Development Lifecycle (SDLC)

Baking security into every phase — from idea to production

2.1 What Is Secure SDLC?

The **Software Development Life Cycle (SDLC)** is the structured process by which software is conceived, designed, developed, tested, deployed, and maintained. While traditional SDLC models focused primarily on functionality, delivery speed, and performance, they often treated security as a separate process, verifying it only after the software was built.

Secure SDLC (SSDLC) shifts this paradigm by embedding security at every phase of the development lifecycle, ensuring that security risks are identified, managed, and mitigated from the very beginning. The goal is to eliminate vulnerabilities early, when they are cheapest to fix, and to create software that is secure by design.

The Traditional SDLC vs. Secure SDLC

Stage	Traditional SDLC Focus	Secure SDLC Focus
Requirements	Functional specs, user needs	Security requirements, compliance, threat modeling

Stage	Traditional SDLC Focus	Secure SDLC Focus
Design	Architecture, system diagrams	Secure design patterns, trust boundaries, data flows
Development	Writing code, feature delivery	Secure coding standards, input validation, code reviews
Testing	Functional & performance tests	Security testing (SAST, DAST, SCA, fuzzing, penetration testing)
Deployment	Packaging & deployment pipelines	Deployment hardening, secure configuration management
Maintenance	Bug fixes, feature updates	Continuous monitoring, patching, vulnerability management

Key Principles of Secure SDLC

- Shift Left Approach: Security activities begin early, during the requirements gathering and design phases.

- Continuous Security Validation: Security checks are automated and integrated into CI/CD pipelines.

- Collaboration: Developers, security engineers, operations, QA, and compliance teams work together.

- Security as a Quality Attribute: Secure software isn't a separate goal; it's part of delivering high-quality software.

- Compliance Built-In: Regulatory requirements (e.g., HIPAA, GDPR, PCI-DSS, SOC 2) are accounted for from the start.

Secure SDLC is not a parallel process; it's a fully integrated part of modern software delivery.

Why Secure SDLC Is Essential in SecDevOps

In modern software ecosystems:

- Code changes are deployed daily or hourly.

- Teams often rely on third-party packages and open-source libraries.

- Applications are deployed across multi-cloud, containerized environments.

- Attackers actively scan for weak entry points introduced during rushed releases.

Secure SDLC ensures that security is not compromised for the sake of speed; instead, it is embedded into the very workflow that enables rapid delivery.

The Cost-Benefit of Secure SDLC

Stage Detected	Relative Cost to Fix Vulnerability
Requirements	1x cost
Design	5x cost
Development	10x cost
Testing	15x cost
Post-Deployment	30x–100x cost

Early detection reduces not only financial cost but also reputational risk and operational disruption.

Secure SDLC in the Context of SecDevOps

While Secure SDLC formalizes when security should happen, SecDevOps focuses on how it is operationalized continuously:

- Secure SDLC defines the framework.

- SecDevOps provides automation, tooling, and cultural alignment to make Secure SDLC work at DevOps speed.

Without Secure SDLC, SecDevOps has no structured foundation. Without SecDevOps, Secure SDLC cannot scale in fast-paced pipelines.

Visual Flow of Secure SDLC

CopyEdit

Requirements → Secure Design → Secure Coding → Security Testing → Secure Deployment → Secure Operations → Continuous Feedback → Continuous Improvement

Each phase feeds into the next, creating an end-to-end security loop.

Key Outcomes of Secure SDLC

- Fewer vulnerabilities reaching production

- Faster security remediation cycles

- Improved regulatory compliance posture

- Lower long-term development costs

- Increased confidence for both developers and customers

2.2 Why Secure SDLC Is Foundational to SecDevOps

At its core, SecDevOps is not a replacement for Secure Software Development Life Cycle (SDLC); it is a natural extension of it. While Secure SDLC defines *what* must happen and *when* in the lifecycle, SecDevOps operationalizes these activities in highly automated, fast-paced, cloud-native, and collaborative environments. A successful SecDevOps program cannot be built without a thorough understanding of Secure Software Development Life Cycle (SDLC) principles. Secure SDLC gives the structure; SecDevOps gives the scale and speed.

How Secure SDLC Provides the Foundation

Secure SDLC Role	SecDevOps Alignment
Defines security checkpoints for each phase	Automates security testing and enforcement
Aligns cross-functional teams on secure processes	Breaks down silos between Dev, Sec, Ops
Establishes security requirements early	Converts those requirements into automated checks
Identifies risks through threat modeling	Automates threat validation via scans, fuzzers, and policies
Builds security into designs	Enforces design rules via IaC scanning and policy-as-code
Prioritizes secure coding practices	Embeds secure coding tools directly into IDEs and pipelines

Secure SDLC = the policy framework; SecDevOps = the delivery mechanism.

The Challenge Without Secure SDLC

Without Secure SDLC principles, even the best SecDevOps pipelines will fail because:

- Security tests may exist but lack context (testing the wrong things).

- Developers may not know how to code securely, leading to recurring vulnerabilities.

- Security may still feel like "somebody else's problem" to engineering teams.

- Security debt will accumulate, creating fragile, hard-to-secure systems.

In other words, SecDevOps without a Secure Software Development Life Cycle (SDLC) becomes tool-driven chaos, rather than adequate security.

Key Reasons Secure SDLC Is Essential Before SecDevOps

1. Security Becomes a Design Attribute

Security is no longer reactive; it's embedded into:

- Design decisions

- Architecture reviews

- Data flow diagrams

- Component-level security

This enables better risk ownership from the outset of the development process.

2. Enables Meaningful Automation

You cannot automate what you don't understand.

Secure SDLC:

- Provides known security policies, rules, and requirements.

- Enables effective automation in pipelines (SAST, DAST, IaC scanning).

- Prevents "false sense of security" from simply running unstructured scanners.

Automation only works well after standards and baselines are defined.

3. Supports Measurable Security Outcomes

Secure SDLC lays the groundwork for:

- Consistent metrics (e.g., vulnerability reduction rate)

- Prioritization of issues based on design-level criticality

- Continuous improvement loops

SecDevOps enables metrics-driven security maturity once Secure SDLC rules are in place.

4. Aligns Security with Business and Compliance

Secure SDLC helps:

- Map regulatory frameworks into specific control requirements.

- Integrate security into business goals.

- Ensure compliance isn't bolted on, but designed in.

Example:
PCI-DSS requires strong access controls → Secure SDLC ensures role definitions exist → SecDevOps enforces these in IaC templates and cloud policies.

Real-World View: The Relationship

If Secure SDLC answers:

- What should we secure?

- When should we secure it?

Then SecDevOps answers:

- How do we secure it continuously?

- How do we enforce security automatically?

- How do we scale security across fast-moving teams?

They are inseparable. Without Secure SDLC, SecDevOps pipelines may run but lack meaningful governance.

Visual Summary

markdown

CopyEdit

Secure SDLC → Defines Standards → Enables Automation → Drives Continuous SecDevOps

Summary Key Takeaways

- Secure SDLC gives SecDevOps its structure, context, and purpose.

- SecDevOps converts Secure SDLC into a continuous, automated practice.

- One without the other leads to either bureaucracy (Secure SDLC only) or chaos (SecDevOps only).

Adequate security requires both strong processes and strong pipelines; Secure SDLC provides the process foundation that SecDevOps executes.

2.3 Secure SDLC Phases and Security Activities

Secure SDLC applies security principles at every stage of the software development lifecycle. Each phase offers unique opportunities to:

- Prevent vulnerabilities before they exist

- Detect weaknesses earlier

- Reduce the cost of remediation

- Embed compliance naturally

SecDevOps operationalizes Secure SDLC by automating many of these activities, but first, we must understand where security fits into each phase.

2.3.1 Phase 1 — Requirements Gathering (Secure Planning)

Security Objective:
Identify security goals, business risks, and compliance obligations early.

Key Activities:

- Capture security requirements alongside functional requirements.

- Conduct regulatory mapping (e.g. PCI-DSS, GDPR, HIPAA, SOC 2).

- Identify data classification (sensitive vs. public).

- Perform preliminary risk assessments.

- Initiate high-level threat modeling.

- Determine who owns security responsibilities.

Typical Deliverables:

- Security Requirements Document (SRD)

- Risk Register

- Data Protection Impact Assessment (DPIA)

- Business Continuity considerations

Starting with security here prevents downstream architectural gaps.

2.3.2 Phase 2 — Secure Design

Security Objective:
Architect systems to minimize attack surfaces and enforce trust boundaries.

Key Activities:

- Threat modeling using STRIDE, PASTA, or LINDDUN frameworks.

- Build Data Flow Diagrams (DFDs) to identify trust boundaries.

- Apply secure design patterns (least privilege, defense-in-depth).

- Determine authentication, authorization, and encryption standards.

- Conduct architectural security reviews.

Typical Deliverables:

- Secure Design Document (SDD)

- Architecture Diagrams with Trust Boundaries

- Encryption Key Management Plan

- Identity & Access Control Models

Design flaws are often cheaper to fix now than after coding starts.

2.3.3 Phase 3 — Secure Development (Coding Phase)

Security Objective:
Write secure code that proactively prevents vulnerabilities.

Key Activities:

- Enforce secure coding standards (language-specific guidelines).

- Use Static Application Security Testing (SAST) tools during coding.

- Integrate IDE security plugins for real-time code analysis.

- Perform peer code reviews with security checklists.

- Use vetted, well-maintained open-source packages.

- Prevent hardcoded credentials via secrets management.

Typical Tools:

- Semgrep, CodeQL, SonarQube (SAST)

- Gitleaks, TruffleHog (Secrets Scanning)

- Dependency Check, Snyk (SCA)

Deliverables:

- Secure Code Repositories

- Code Review Logs

- Approved Dependency Lists (SBOM)

Secure coding isn't about perfection; it's about repeatable hygiene.

2.3.4 Phase 4 — Secure Testing

Security Objective:
Validate that code and infrastructure behave securely in runtime conditions.

Key Activities:

- Run Dynamic Application Security Testing (DAST).

- Perform API Security Testing (OWASP API Top 10).

- Conduct Interactive Application Security Testing (IAST).

- Apply fuzz testing for unpredictable input scenarios.

- Scan Infrastructure-as-Code templates.

- Execute container image scans before deployment.

- Perform penetration tests (automated and manual).

Typical Tools:

- OWASP ZAP, Burp Suite (DAST)

- Checkov, Terrascan (IaC Scanning)

- Trivy, Aqua, Clair (Container Scanning)

- Postman, SoapUI (API Testing)

- JFuzz, AFL (Fuzz Testing)

Deliverables:

- Security Test Reports

- Vulnerability Findings & Remediation Plans

- Passed Build Artifacts

Security testing should be integrated directly into the CI/CD pipeline, rather than being left to manual testing teams alone.

2.3.5 Phase 5 — Secure Deployment

Security Objective:
Ensure that deployment pipelines enforce consistent, secure releases.

Key Activities:

- Implement artifact signing and verification.

- Scan images and code before deployment (final security gate).

- Use policy-as-code to enforce deployment constraints.

- Limit permissions and keys for deployment automation.

- Monitor runtime configurations for deviations.

- Validate deployment templates (e.g., Helm Charts, Terraform).

Typical Tools:

- Sigstore, Cosign (Artifact Signing)

- OPA/Gatekeeper, Kyverno (Policy Enforcement)

- HashiCorp Vault (Secrets Injection)

- ArgoCD, Flux (GitOps Deployment Security)

Deliverables:

- Signed Build Artifacts

- Policy Validation Logs

- Deployment Audit Trails

Deployment security is not only about initial safety, but also about repeatable, predictable releases.

2.3.6 Phase 6 — Secure Operations (Post-Deployment Monitoring)

Security Objective:
Continuously monitor systems for emerging threats, misconfigurations, and active attacks.

Key Activities:

- Implement centralized security logging and monitoring.

- Set up intrusion detection and anomaly monitoring.

- Enable runtime security for containers and serverless workloads.

- Enforce vulnerability management for live infrastructure.

- Regularly review IAM policies and permissions drift.

- Integrate security incident response playbooks into operations.

- Periodically rehearse incident response drills.

Typical Tools:

- Falco, Sysdig, Aqua Trivy (Runtime Security)

- ELK, Splunk, Datadog, SIEM Platforms (Centralized Monitoring)

- AWS Security Hub, Azure Security Center (Cloud Security Posture)

- PagerDuty, OpsGenie (Alerting & Incident Response)

Deliverables:

- Security Incident Logs

- Weekly Security Health Reports

- Updated IR Playbooks

Operations must assume systems will fail; resilience depends on fast detection and response.

2.3.7 The Continuous Feedback Loop

SecDevOps thrives on feedback loops. Lessons learned in operations inform better requirements, design, and development.

Stage	Feedback Examples
Operations ➔ Requirements	Post-incident findings lead to new security stories
Testing ➔ Development	Vulnerability findings improve coding standards
Deployment ➔ Design	Misconfiguration patterns inform better templates
Threat Modeling ➔ Planning	Emerging threats update security requirements

Security maturity is not static; it's iterative.

Summary Takeaway

Secure SDLC Phase	SecDevOps Transformation
Requirements	Threat Modeling-as-Code, Policy-as-Code
Design	Automated Architecture Validation
Development	Secure Coding Tools & Pipelines

Secure SDLC Phase	SecDevOps Transformation
Testing	Continuous Security Testing
Deployment	Secure CI/CD Controls
Operations	Runtime Monitoring & Automated IR

By tightly coupling Secure SDLC and SecDevOps, organizations establish a seamless chain of security ownership that extends from planning to production.

2.4 Challenges in Implementing Secure SDLC in Fast-Paced DevOps Environments

While Secure SDLC provides the blueprint for building secure software, implementing it within modern, fast-moving DevOps environments presents significant challenges.

Agile teams prioritize rapid iteration, continuous deployment, and minimal disruption, which can easily create tension between speed and security if not managed carefully. Understanding these challenges is the first step toward overcoming them in a successful SecDevOps transformation.

2.4.1 Speed vs. Thoroughness

The Conflict:

DevOps thrives on rapid delivery. Secure SDLC demands careful analysis, risk assessments, and formalized reviews.

Teams fear that security may slow down delivery pipelines or introduce bureaucracy that disrupts business agility.

Symptoms:

- Security reviews postponed or skipped entirely.

- Threat modeling is seen as "nice to have" instead of essential.

- Last-minute security sign-offs bottleneck releases.

Resolution Strategies:

- Break security tasks into incremental work items that fit agile sprints.

- Use "security user stories" as part of product backlog grooming.

- Build automated gates into CI/CD pipelines to prevent human bottlenecks.

- Practice lightweight, continuous threat modeling instead of one-time heavyweight reviews.

Security work must scale with the team's delivery speed, not oppose it.

2.4.2 Lack of Security Expertise Among Developers

The Conflict:
Many developers receive little or no formal training in secure coding or threat modeling. As a result, insecure design decisions are often made unknowingly, creating technical debt that compounds over time.

Symptoms:

- Repeated reintroduction of common vulnerabilities (e.g., SQL injection, hardcoded secrets).

- Insecure third-party libraries chosen due to convenience.

- Fear or confusion when asked to implement security controls.

Resolution Strategies:

- Launch developer security enablement programs (bootcamps, microlearning, certifications).

- Build security champion networks — developers with special training embedded within each team.

- Offer secure code review office hours led by security engineers.

- Incorporate security metrics into developer KPIs (e.g., % of PRs passing SAST).

Security skills must be treated as core competencies for modern software engineering.

2.4.3 Tool Overload and Poor Tool Integration

The Conflict:
As organizations adopt more security tools, developers may face a fragmented toolchain with:

- Overlapping scanners

- Conflicting results

- Unclear ownership of remediation

This leads to alert fatigue, analysis paralysis, and a lack of confidence in findings.

Symptoms:

- Duplicate vulnerabilities reported by multiple tools.

- Different teams are relying on conflicting dashboards.

- Inconsistent enforcement of security gates.

Resolution Strategies:

- Standardize on a centralized security toolchain integrated directly into the CI/CD pipeline.

- Aggregate findings into a single pane of glass (e.g., unified dashboards or GRC platforms).

- Use orchestration platforms (e.g., AppSec orchestration) to coordinate multiple scanning tools.

- Defining ownership models: who triages, who fixes, who tracks.

The goal is seamless security integration, not "yet another tool" in the developer's workflow.

2.4.4 Cultural Resistance

The Conflict:
Security has historically been viewed as the "department of no," leading to distrust or avoidance by engineering teams. Cultural inertia creates resistance to changing workflows, especially when security is perceived as:

- Slowing innovation

- Adding bureaucracy

- Not understanding real engineering pressures

Symptoms:

- Security tickets are regularly deprioritized.

- Passive-aggressive workarounds to bypass controls.

- "It's not my problem" mindset among developers.

Resolution Strategies:

- Shift from "compliance policing" to "security as partnership."

- Involve developers directly in threat modeling and control design.

- Create leadership-aligned mandates for security of ownership at every level.

- Recognize and reward security-positive behaviors.

- Use blameless postmortems to turn incidents into learning opportunities.

Security must embed itself as a service to developers, not as an enforcer above them.

2.4.5 Dynamic, Cloud-Native Environments

The Conflict:

Modern cloud-native systems are:

- Highly dynamic

- Composed of short-lived resources

- Deployed across multi-cloud architectures

- Heavily reliant on third-party APIs

Traditional security models, built for static servers, struggle to adapt. Controls must evolve to handle ephemeral, scalable infrastructure.

Symptoms:

- Limited visibility into runtime cloud configurations.

- Lack of consistent configuration baselines across environments.

- Manual audits are unable to keep pace with cloud resource changes.

Resolution Strategies:

- Adopt Infrastructure-as-Code (IaC) to ensure consistent deployment definitions.

- Implement IaC scanning tools (Checkov, Terrascan) before provisioning.

- Use Cloud Security Posture Management (CSPM) platforms for continuous cloud compliance.

- Deploy runtime monitoring for containerized workloads (Falco, Sysdig).

Security must match the pace and flexibility of dynamic infrastructure, without losing control.

2.4.6 Shifting Threat Landscape

The Conflict:
Attackers continually adapt their techniques, targeting:

- CI/CD pipelines themselves

- Supply chains and open-source dependencies

- Credential leaks

- Build servers and orchestrators

A Secure SDLC must account for emerging threats, not just legacy ones.

Symptoms:

- Third-party compromise impacting internal builds.

- Dependency confusion attacks injecting malicious libraries.

- Exposed access tokens or API keys in public repos.

Resolution Strategies:

- Adopt Software Bill of Materials (SBOMs) to track dependencies.

- Sign build artifacts (e.g., using Sigstore, Cosign).

- Scan open-source packages regularly (e.g., Snyk, OWASP Dependency-Check).

- Implement secrets scanning in pre-commit hooks (e.g., Gitleaks, TruffleHog).

Security isn't static; threat modeling and tooling must continuously evolve with new risks.

2.4.7 Summary Table of Challenges & Solutions

Challenge	Key Resolution Techniques
Speed vs. Thoroughness	Incremental modeling, security user stories, CI/CD automation
Lack of Developer Expertise	Security champions, developer training, secure coding KPIs
Tool Overload	Toolchain consolidation, centralized dashboards, orchestration
Cultural Resistance	Partnership mindset, leadership support, and positive reinforcement
Dynamic Environments	IaC scanning, CSPM, runtime container security
Shifting Threats	SBOMs, signed artifacts, supply chain scanning, secrets scanning

Final Takeaway

While the technical aspects of Secure SDLC are well-defined, its successful implementation requires organizations to overcome cultural, technical, educational, and organizational challenges simultaneously. Overcoming these challenges lays the proper foundation for a mature, fully automated, and scalable Security Development and Operations (SecDevOps) program.

Chapter 3: Securing Code Repositories and Git Workflows

Protecting the source: integrity, trust, and secure collaboration

3.1 Why Securing Code Repositories Matters

Modern software development relies heavily on distributed version control systems, such as Git, and centralized platforms, including GitHub, GitLab, and Bitbucket. These repositories are more than just collections of code; they are mission-critical assets that contain:

- Proprietary source code

- API keys and secrets

- Deployment configurations

- Infrastructure-as-Code files

- CI/CD workflows

- Access metadata

Because of their central role in the software supply chain, code repositories are high-value targets for attackers. A single compromise can cascade across build pipelines, cloud environments, and end-user systems. When source control is breached, the consequences can include code theft, malware injection, backdoors, and unauthorized access to infrastructure. Securing Git workflow ensures integrity, traceability, and trust in your software supply chain.

3.2 Git Security Best Practices

While Git is a powerful and flexible version control system, it was not designed with security as a default priority. Misconfigurations, poor developer hygiene, and unregulated access can quickly turn a Git repository into a vulnerability gateway. To build secure development environments, teams must adopt structured and enforceable Git security practices. These should cover both local developer workflows and server-side repository management to prevent unauthorized code changes, credential exposure, and policy bypasses. The following table summarizes key Git security practices, categorized by focus area, to help teams fortify their repositories and reduce risk across the development lifecycle:

Practice	Why It Matters
Enable multi-factor authentication	Prevents unauthorized Git access
Require signed commits	Ensures commit authenticity
Use branch protection rules	Prevent force-pushes or unreviewed merges
Rotate **SSH/GPG keys** regularly	Avoid stale or compromised access
Limit **admin permissions**	Enforce least privilege in teams

3.3 Secure Git Workflows

A secure repository alone is not enough; the way developers interact with Git plays a critical role in maintaining code integrity and traceability. Poor workflows can lead to accidental commits of secrets, unreviewed code merges, or even bypass of security controls. To address these risks, teams should establish and enforce secure Git workflows that are consistent, auditable, and automation-friendly. These workflows must strike a balance between developer velocity and policy enforcement, ensuring that security is not compromised for the sake of convenience. Below are recommended secure Git workflow practices that help teams embed security from the moment code is written through to production deployment:

Recommended Practices:

- Create protected branches (e.g., main, release)

- Use Pull Requests (PRs) for all changes

- Require at least one reviewer before merge

- Disallow direct pushes to protected branches

- Enforce CI checks before merging

Example GitHub Settings:

yaml

CopyEdit

branches:

 - name: main

protection:

  required_status_checks:

   - lint

   - test

   - scan

  enforce_admins: true

  required_pull_request_reviews:

   required_approving_review_count: 1

3.4 Secrets in Code — A Hidden Threat

Hardcoded secrets, such as API keys, database passwords, cloud credentials, and encryption tokens, are one of the most common and dangerous security mistakes in modern software development. When these secrets are embedded in code and pushed to repositories, they create silent vulnerabilities that attackers can easily exploit.

In public or even private repositories, secrets can be:

- Scanned by bots within minutes of exposure

- Harvested from Git history, even after removal

- Used to pivot into production environments and third-party services

Because secrets often carry elevated privileges, a single exposed key can lead to complete system compromise, data theft, or infrastructure abuse, often without immediate detection. To combat this threat, organizations must adopt proactive detection, prevention, and rotation strategies, as outlined in the next section. Over 85% of breaches involve leaked credentials. Secrets (API keys, tokens, passwords) accidentally committed to Git constitute a significant risk.

Common Mistakes:

- Hardcoded secrets in .env, .py, .yaml, or config.js

- Pushed SSH keys or JWT tokens

- Committing .env to repo

Prevention:

- Add *.env, secrets.yml to .gitignore

- Use Git hooks to block commits with secrets

- Scan with tools like:

    o Gitleaks

    o TruffleHog

    o GitGuardian

3.5 Git Hooks and Pre-Commit Security

Git hooks provide a powerful yet often underutilized mechanism for enforcing security and quality checks directly within the developer workflow. These are customizable scripts that Git executes at defined points in the version control lifecycle, such as before committing or pushing code.

Pre-commit hooks allow teams to catch problems *before* code ever leaves a developer's machine. This early interception is crucial for:

- Preventing accidental commits of secrets or large binaries

- Enforcing secure code formatting or linting

- Validating commit messages for traceability

- Running lightweight static analysis or vulnerability scans

By leveraging Git hooks, organizations can shift security left, turning every developer action into an opportunity to enforce secure, standardized, and policy-aligned behavior. Below are key practices and tools that enhance security through Git pre-commit automation:

Use Git hooks to enforce local rules before a developer even pushes code.

Example: Block commits with secrets

bash

CopyEdit

#!/bin/bash

if grep -r 'AWS_SECRET' .; then

   echo "Secret detected. Commit aborted."

   exit 1

fi

Automate with Pre-Commit Framework:

yaml

CopyEdit

repos:

  - repo: https://github.com/pre-commit/pre-commit-hooks

    hooks:

      - id: check-yaml

      - id: check-merge-conflict

      - id: detect-aws-credentials

3.6 Git Commit Signing (GPG or SSH)

In modern software supply chains, verifying the authenticity and integrity of code contributions is essential. One of the most effective ways to achieve this is through Git commit signing, which cryptographically verifies that a commit or tag came from a trusted developer. By signing commits using GPG keys (GNU Privacy Guard) or SSH keys, developers create a tamper-evident chain of custody. This not only protects against impersonation attacks and unauthorized changes but also strengthens trust in open-source and enterprise projects.

Signed commits help:

- Ensure code provenance in multi-author repositories

- Detect unauthorized contributors or compromised accounts

- Meet compliance standards (e.g., SLSA, NIST SSDF, supply chain security)

- Enhance traceability across regulated or critical environments

The following section explains how commit signing works, compares GPG vs SSH methods, and outlines steps to integrate signature enforcement into Git workflows.

How Git Commit Signing Works

When you sign a Git commit or tag, Git appends a cryptographic signature to the metadata of the commit. This signature is created using your private key (GPG or SSH), and anyone with your corresponding public key can verify:

- That you authored the commit, and

- That the content of the commit has not been altered since signing.

Signed commits are marked as "verified" by platforms like GitHub, GitLab, and Bitbucket, providing a clear signal of authenticity and trust.

GPG vs. SSH for Commit Signing

Feature	GPG (GNU Privacy Guard)	SSH (Secure Shell Keys)
Primary Use	Commit/tag signing and general cryptography	Authentication for Git, servers, commit signing
Key Type	OpenPGP format (RSA, Ed25519, etc.)	SSH public/private keypair
Tooling Complexity	Higher (manual setup, key servers)	Lower (leverages existing SSH keys)
GitHub Support	Full support, requires key upload to GitHub	Full support, seamless for developers with SSH
Verification Process	Based on Web of Trust or key server fingerprints	Tied to your GitHub account and key fingerprint
Best Use Case	Highly secure environments with GPG policies	Simpler setups, especially where SSH is already used

Both are acceptable. SSH signing is now more common due to ease of use, but GPG is still preferred in some regulated industries and enterprise workflows.

Setting Up Commit Signing

For GPG:

1. **Generate a GPG key***:*

bash

CopyEdit

gpg --full-generate-key

2. **List keys and copy your key ID***:*

bash

CopyEdit

gpg --list-secret-keys --keyid-format=long

3. **Configure Git** *to sign commits:*

bash

CopyEdit

git config --global user.signingkey <YOUR_KEY_ID>

git config --global commit.gpgsign true

4. **Upload your public key** *to GitHub/GitLab/Bitbucket*

For SSH Signing (Git ≥ 2.34):

1. **Generate a new SSH key** *(if needed):*

bash

CopyEdit

ssh-keygen -t ed25519 -C "your_email@example.com"

2. ***Enable SSH signing in Git****:*

bash

CopyEdit

git config --global gpg.format ssh

git config --global user.signingkey ~/.ssh/id_ed25519.pub

git config --global commit.gpgsign true

3. ***Add SSH public key*** *to your Git provider's settings*

Enforcing Commit Signature Verification

Organizations can require signed commits by:

- Enabling "Require signed commits" in GitHub/GitLab branch protection rules

- Adding CI checks to reject unsigned commits in PRs

- Defining commit signature policies in developer handbooks

Best Practices Summary

- Always sign commits and tags, especially for open-source/public projects

- Store private keys securely (consider password-protected or hardware-backed keys)

- Rotate keys periodically and revoke compromised keys immediately

- Combine signing with identity verification tools (e.g., SLSA, Sigstore)

Signed commits verify who made a change, protecting against impersonation.

- Use GPG or SSH keys

- GitHub will show a green "Verified" badge

- Require verified commits via branch settings

bash

CopyEdit

git config --global user.signingkey YOURKEYID

git config --global commit.gpgsign true

3.7 Real-World Breach Example

Understanding the real-world impact of insecure code repositories is crucial for appreciating the importance of SecDevOps practices. While theoretical threats help define risk, actual breaches reveal the tangible consequences of overlooked security controls, ranging from code exfiltration to unauthorized infrastructure access and damage to reputation.

In this section, we examine a high-profile incident in which attackers exploited weak Git hygiene and repository misconfigurations to compromise a software supply chain. This example illustrates how:

- Exposed secrets led to lateral movement

- Poor access control allowed repository manipulation

- Lack of commit signing obscured accountability

- Absence of automated scanning delays detection

By analyzing what went wrong and how it could have been prevented, we provide actionable lessons that organizations can apply to secure their development workflows.

In 2022, Heroku and TravisCI had OAuth tokens exposed in public GitHub repositories. Attackers leveraged these tokens to pivot into CI pipelines and extract secrets.

Lesson: Enforce secret scanning, use least privilege OAuth scopes, and monitor public forks.

3.8 Chapter Summary

Securing the codebase and Git repositories is one of the foundational pillars of a robust SecDevOps strategy. As we've seen throughout this chapter, the way source code is stored, shared, and modified can either strengthen or weaken an organization's overall security posture.

Key takeaways include:

- Git repositories are not just code; they contain secrets, configurations, and infrastructure artifacts that can be exploited.

- Secure Git workflows and policies must include commit signing, secrets scanning, branch protection, and access control.

- Pre-commit hooks and CI-based scanners provide proactive enforcement and prevent risky code from entering the main codebase.

- Real-world breaches reveal that minor oversights, like hardcoded keys or unsigned commits, can lead to massive downstream compromise.

By treating Git repositories as critical infrastructure, teams can reduce their attack surface and set a strong foundation for secure, compliant, and high-integrity software development.

Part II: Application and Infrastructure Security in the Pipeline

This section explores the automation and proactive scanning of code, dependencies, containers, and infrastructure.

- Chapter 4: Static and Dynamic Code Analysis (SAST & DAST)
  Covers automated code scanning tools, best practices, and integrating them into CI/CD.
- Chapter 5: Dependency and Container Security
  Explains how to secure open-source components, perform software composition analysis (SCA), and harden container images.
- Chapter 6: Infrastructure as Code (IaC) Security
  Addresses how to secure declarative infrastructure, detect misconfigurations, and integrate IaC scanning tools.
- Chapter 7: Secrets Management and Identity Security
  Provides strategies for securing secrets, managing identities, and implementing least privilege access in automated environments.

Chapter 4 — Static and Dynamic Code Analysis (SAST & DAST)

"Automation catches what human eyes miss. The right scanners make your pipelines your first firewall."

Code scanning tools are one of the core pillars of automation in Secure SDLC and SecDevOps. In a secure development lifecycle, identifying vulnerabilities early, before code reaches production, is not just recommended, it's essential. Two of the most effective ways to do this are through Static Application Security Testing (SAST) and Dynamic Application Security Testing (DAST). These two approaches work at different stages of the SDLC:

- SAST scans code for weaknesses before it's ever run.

- DAST tests the application while it's live and executing.

Used together, they provide defense in depth, catching different types of flaws that could otherwise slip through unnoticed. This chapter examines the role of SAST and DAST within modern CI/CD pipelines, provides guidance on interpreting their findings, and outlines strategies for integrating popular tools such as SonarQube, Semgrep, and OWASP ZAP into automated workflows. Whether you're a developer, security engineer, or DevOps lead, mastering code analysis will help you build resilient software from the inside out.

Static and dynamic analysis allow teams to:

- Catch vulnerabilities early.

- Shift security left into development pipelines.

- Build repeatable, automated assurance processes.

- Provide instant feedback to developers.

In this chapter, we'll explore how SAST and DAST integrate into CI/CD pipelines, how they work, how to interpret results, and how to tune them for maximum value.

4.1 What Are SAST and DAST?

Security vulnerabilities can originate from coding mistakes, misconfigurations, or risky application behavior. To detect these flaws effectively, developers and security teams use two complementary approaches: Static Application Security Testing (SAST) and Dynamic Application Security Testing (DAST). SAST analyzes the source code, bytecode, or binaries without executing the program. It's designed to detect security weaknesses, such as hardcoded secrets, insecure functions, or logic flaws, early in the development process. DAST, in contrast, tests a running application from the outside, simulating real-world attacks. It identifies runtime vulnerabilities, such as SQL injection, authentication bypasses, and insecure session handling, which SAST often misses. Both techniques have their strengths and limitations. Used together, they create a comprehensive security testing strategy across the entire software lifecycle.

The following table outlines the key differences between SAST and DAST:

Type	What It Analyzes	When It Runs
SAST (Static Application Security Testing)	Source code, bytecode, or binaries	During development, pre-build
DAST (Dynamic Application Security Testing)	Running applications (black-box tests)	After deployment to test/staging environments

SAST sees vulnerabilities in logic before code runs. DAST sees vulnerabilities in behavior after code runs.

Why You Need Both

While SAST and DAST serve different purposes and operate at different stages of the software development lifecycle, neither is a complete solution on its own. Relying solely on one can leave critical gaps in application security coverage.

- SAST is excellent for identifying vulnerabilities in the codebase early, but it may miss runtime behaviors or configuration-related issues that only appear during execution.

- DAST, on the other hand, uncovers vulnerabilities during application runtime, but it lacks insight into the internal logic, source code, or development practices.

By combining both approaches, teams can achieve end-to-end visibility across the development and deployment pipeline, identifying vulnerabilities both before and during the application's execution. The table below compares the strengths of each and illustrates how they complement one another in a secure DevOps workflow:

Vulnerability Type	SAST Best At	DAST Best At
Input validation errors	✓	✓
Insecure crypto implementations	✓	x
Sensitive data exposure	✓	✓
Authentication flaws	✓	✓
Authorization bypasses	x	✓
Logic bugs in APIs	x	✓
Server misconfiguration	x	✓

Together, SAST + DAST provide full coverage across code logic AND deployed runtime behavior.

4.2 Where SAST and DAST Fit into CI/CD Pipelines

In a SecDevOps environment, automation is key, and that includes security testing. To be effective, SAST and DAST must be tightly integrated into continuous Integration/Continuous Deployment (CI/CD) pipelines, rather than being left as optional, manual steps.

Embedding these tools ensures that:

- Vulnerabilities are detected and remediated early and often

- Risk assessments are consistent and automated

- Developers receive immediate feedback on the security impact of their changes

- Security becomes a shared responsibility, not a late-stage bottleneck

SAST and DAST can be strategically placed at different phases of the pipeline to maximize their strengths. The sections below explain where and how to embed each type of testing into your DevOps workflow, from code commit to production deployment.

SAST Integration Points:

- Pre-commit hooks (locally)

- Pull request triggers

- CI build pipelines

DAST Integration Points:

- Post-deployment to test or staging environments

- Automated nightly security test suites

- Pre-production gates before releases

Typical CI/CD Flow:

plaintext

Developer → *Commit* → *SAST* → *Build* → *Deploy* →
DAST → *Release Approval* → *Production*

Automated security gates block insecure code from progressing.

4.3 Popular SAST Tools and Integrations

Selecting the right Static Application Security Testing (SAST) tool is essential for embedding secure coding practices into the development pipeline. A good SAST tool should not only identify vulnerabilities in real-time but also integrate seamlessly into your CI/CD workflows, provide actionable feedback to developers, and support the languages and frameworks the team uses.

Many modern SAST solutions support features like:

- IDE integration for inline feedback

- GitHub/GitLab pre-merge scanning

- CI pipeline hooks (e.g., Jenkins, Azure DevOps, GitHub Actions)

- Custom rule creation and policy enforcement

- Developer-friendly output formats and integrations with ticketing systems

Below is a list of popular and widely adopted SAST tools, along with their integration options and common use cases in SecDevOps environments:

Tool	Languages Supported	Notes
Semgrep	Python, JavaScript, Go, Java, many others	Lightweight, developer-friendly, highly tunable
CodeQL (GitHub Advanced Security)	C, C++, Java, JavaScript, C#, Python	Query-based vulnerability detection
SonarQube	Wide multi-language support	Quality + security scanning combined
Checkmarx, Fortify, Veracode	Enterprise SAST suites	Deeper coverage with complex integrations

Example: Semgrep in CI Pipeline

GitHub Action Example:

yaml

CopyEdit

name: SAST with Semgrep

on: pull_request

jobs:

* semgrep:*

*   runs-on: ubuntu-latest*

steps:

   - uses: actions/checkout@v2

   - uses: returntocorp/semgrep-action@v1

   with:

   config: 'auto'

Semgrep provides instant pull request (PR) feedback with actionable results.

4.4 Popular DAST Tools and Integrations

Dynamic Application Security Testing (DAST) tools play a vital role in detecting runtime vulnerabilities that static analysis may miss. These tools simulate external attacks against a running application, mimicking how real-world attackers probe for weaknesses like SQL injection, cross-site scripting (XSS), insecure session handling, and misconfigured headers.

Modern DAST tools are built to:

- Perform black-box testing without needing access to source code

- Integrate into CI/CD pipelines to run after deployment to staging environments

- Provide actionable remediation guidance for developers

- Support authenticated scans and API testing

- Offer automated crawling and fuzzing to uncover hidden flaws

Below is a list of widely used DAST tools, along with their capabilities and integration methods within a SecDevOps pipeline:

Tool	Focus Area	Integration Method
OWASP ZAP	Web app vulnerability scanner	API integrations, CI/CD compatible
Burp Suite (Pro)	Deep web & API fuzzing	Manual + automated tests
Arachni	Ruby-based scanner	REST APIs, web apps
Nikto	Lightweight server scanner	CLI automation
Postman/Newman Security Collections	API endpoint validation	Automated API security tests

Example: OWASP ZAP in CI

Dockerized Pipeline Example:

bash

CopyEdit

docker run -v $(pwd):/zap/wrk/:rw owasp/zap2docker-stable zap-baseline.py -t https://staging-app-url.com

Easily inserted into build pipelines for automated staging environment scans.

4.5 Interpreting SAST & DAST Findings

Running SAST and DAST tools is only the first step; knowing how to interpret their results is what turns raw data into actionable security improvements. These tools often generate large volumes of findings, including false positives, duplicate alerts, and low-risk warnings. Without proper understanding, teams may experience alert fatigue or overlook critical issues.

To effectively triage and respond to security test results, teams must:

- Understand the risk categories reported (e.g., critical, high, medium, low)

- Identify true positives vs. false alarms

- Correlate findings with the codebase or application behavior

- Prioritize issues based on exploitability and business impact

- Integrate results into ticketing and remediation workflows

The following sections summarize typical findings produced by SAST and DAST tools, giving a practical reference to help distinguish signal from noise and drive targeted improvements.

SAST Typical Findings:

- SQL Injection risks

- Cross-site scripting (XSS)

- Command injection

- Path traversal

- Sensitive data exposure

- Hardcoded secrets

DAST Typical Findings:

- Broken authentication

- Session fixation

- Server misconfiguration

- Authorization flaws

- Sensitive data leakage via APIs

- Input field fuzzing failures

Developers must understand and act on both classes of findings.

4.6 Tuning Scanners to Reduce Noise

One of the most common challenges team faces when implementing SAST and DAST tools is alert fatigue, which occurs when the overwhelming volume of low-priority or irrelevant findings obscures meaningful vulnerabilities. Without tuning, scanners may flood pipelines with false positives, duplicate issues, or non-exploitable warnings, slowing down developers and creating mistrust in security tooling. To make security scanning more effective and developer-friendly, it's essential to fine-tune scanners so they produce high-confidence, actionable results. Tuning may include:

- Customizing rule sets to match the application's risk profile

- Suppressing known false positives or accepted risks

- Adjusting severity thresholds for blocking builds

- Creating context-aware rules based on the codebase or frameworks

- Scheduling deeper scans less frequently, and lighter scans on every commit

Properly tuned scanners align security enforcement with real-world risks, reducing friction between development and security teams while still maintaining strong protection.

One of the most common complaints about SAST and DAST is the occurrence of false positives.

Why False Positives Happen:

- Generic rule sets are not aligned to the codebase.

- Legacy code creates irrelevant warnings.

- Insufficient rule customization.

- Lack of context (e.g., how input is sanitized later).

False Positive Mitigation Techniques:

Approach	Description
Custom Rule Tuning	Build custom Semgrep or CodeQL queries
Baseline Suppression	Ignore known-safe legacy warnings
Taint Analysis	Trace input through code flows
Context-Aware Scanning	Annotate sanitized variables
Scan Scoping	Target only changed code in PR pipelines

Your scanner must know how YOUR team writes code.

4.7 Prioritization: Not All Findings Are Equal

When SAST and DAST tools surface dozens or even hundreds of issues, not all findings demand the same level of attention. Treating every vulnerability as equally critical can paralyze teams, delay releases, or cause security warnings to be ignored altogether.

To manage findings effectively, teams must adopt a risk-based prioritization strategy, one that considers both the technical severity and the business impact of each vulnerability. For example:

- A critical SQL injection in a public API must be addressed immediately

71

- A medium-severity issue in an internal admin tool may be scheduled for the next sprint

- A low-risk finding in a test environment may be deferred or ignored

Prioritization helps focus limited resources on what matters most: preventing real-world exploitation. The table below provides a framework for categorizing vulnerabilities by risk level, urgency, and recommended response time.

Severity	Example Finding	Response
Critical	Hardcoded root credentials	Block merge immediately
High	SQL injection	Prioritize fix in current sprint
Medium	Excessive logging of sensitive data	Scheduled remediation
Low	Unused imports or deprecated APIs	Fix opportunistically

Security debt triaging balances delivery speed and risk reduction.

4.8 Build Security Gates (Fail Conditions)

In a SecDevOps environment, it's not enough to detect vulnerabilities; teams must also enforce security policies automatically. That's where building security gates come in. These are conditional rules embedded into your CI/CD pipelines that determine whether a build should pass or fail based on the results of security scans.

Security gates ensure that:

- Critical or high-severity vulnerabilities never reach production

- Developers are held accountable for fixing issues early

- Security policies are codified and consistent

- Compliance requirements (e.g., SOC 2, ISO 27001) are automatically enforced

By defining fail conditions for SAST, DAST, dependency checks, or container scans, teams can create automated barriers that prevent insecure code from progressing without requiring manual intervention. The table below outlines common failing conditions and how they can be implemented in various stages of the CI/CD process:

Automation enables blocking builds that fail security thresholds.

Gate Type	Trigger Condition
Merge Blocker	Any Critical or High findings
Build Blocker	Dependency vulnerabilities above CVSS 7.0
Artifact Signing Failure	Unsigned build artifacts

Gate Type	Trigger Condition
Secrets Blocker	Any new detected secrets

Blocking gates enforce discipline without relying on human vigilance.

4.9 Combining SAST, DAST & Other Pipelines

While SAST and DAST each play a critical role in securing the software development lifecycle, true resilience comes from integrating multiple layers of security into your CI/CD pipelines. In a mature SecDevOps setup, these tools don't work in isolation; they are combined with dependency scanning, container security, secrets detection, and infrastructure-as-code validation to form a comprehensive, automated security workflow.

By orchestrating these tools together, teams can:

- Detect vulnerabilities at different depths and stages

- Catch misconfigurations and supply chain risks alongside code issues

- Apply consistent security policies across applications, infrastructure, and runtime

- Generate a unified security report per build or deployment cycle

This section outlines how to combine these tools effectively and ensure they operate in parallel, sequentially, or conditionally within your CI/CD pipelines, depending on risk, performance, and compliance needs.

A mature CI/CD pipeline integrates security tools holistically:

CopyEdit

Commit ➜

Secrets Scanning ➜

SAST ➜

Dependency SCA ➜

Build & Artifact Signing ➜

IaC Scanning ➜

Deploy to Staging ➜

DAST ➜

Policy Validation ➜

Manual Review (if needed) ➜

Production Deploy

The pipeline itself becomes a continuously running security control plane.

4.10 Metrics to Measure SAST & DAST Adoption

Implementing SAST and DAST tools is only valuable if their usage leads to measurable improvements in code quality, risk reduction, and team behavior. To ensure that security testing is not just a checkbox, organizations must define and track key metrics that reflect the health, adoption, and effectiveness of these tools across teams and pipelines.

These metrics help answer questions such as:

- Are developers using the tools consistently and correctly?

- Are vulnerabilities being fixed quickly?

- Is secure coding becoming part of the team's culture?

- Are we improving over time?

By collecting actionable data, security leaders can make informed decisions, justify investments, and communicate progress to stakeholders. Below are key metrics to track to evaluate the success of SAST and DAST implementations.

Metric	Meaning
% of PRs scanned with SAST	Coverage discipline
MTTR for security findings	Developer responsiveness
False positive suppression rate	Scanner tuning effectiveness
Critical findings blocked in CI	Prevention strength
Vulnerabilities caught before production	Shift-left maturity

4.11 Final Takeaway

The earlier security issues are detected, the cheaper and safer they are to fix.

SAST and DAST are essential layers of defense, but only when:

- Properly integrated into pipelines.

- Tuned to codebase.

- Coupled with clear ownership and prioritization.

SecDevOps doesn't eliminate vulnerabilities; it builds resilient systems where vulnerabilities are caught early, fixed fast, and prevented systematically.

Chapter 5 — Dependency and Container Security

"You're not just securing your code — you're securing every package, every image, and every external component you trust."

Modern software is no longer built from scratch.
Up to 80–90% of application code today is third-party libraries, packages, APIs, containers, and dependencies. This creates a massive and often invisible attack surface, where supply chain attacks, dependency vulnerabilities, and insecure containers become significant business risks.

In this chapter, we will explore:

- Software Composition Analysis (SCA)

- Software Bill of Materials (SBOM)

- Container image scanning

- Practical tooling

- Automated integration into SecDevOps pipelines

5.1 Dependency Security – Software Composition Analysis

Dependency Security (a.k.a. Software Composition Analysis) refers to scanning, monitoring, and securing the third-party packages, libraries, plugins, and APIs that an application depends on.

If one package has a vulnerability, your application inherits that risk.

Why Dependencies Create Risk:

- Rapid package adoption with minimal vetting.

- Open-source packages may not be actively maintained.

- Attackers increasingly target upstream open-source projects.

- A vulnerability in one package may impact hundreds of companies (e.g. Log4Shell, SolarWinds, LeftPad).

Software Composition Analysis (SCA) is the automated process of:

- Identifying all open-source components used.

- Mapping those components against known vulnerabilities (CVEs).

- Continuously monitoring new vulnerabilities as they emerge.

- Providing visibility into software supply chain risk.

Types of Dependencies Covered by SCA:

Layer	Examples
Language Packages	npm, pip, Maven, NuGet, RubyGems
Containers	Docker Hub base images

Layer	Examples
IaC Modules	Terraform Registry
OS Packages	Alpine, Ubuntu, Amazon Linux repos
APIs	External SaaS dependencies

5.2 SBOM (Software Bill of Materials

A Software Bill of Materials (SBOM) is a machine-readable inventory list that describes all components, including both direct and transitive dependencies, that are included in a software build.

SBOMs allow:

- Full supply chain transparency.

- Faster vulnerability triage when CVEs are announced.

- Compliance with regulatory frameworks (NIST, Executive Orders, ISO standards).

SBOM Formats:

Format	Description
SPDX	Linux Foundation standard
CycloneDX	OWASP-driven format (widely adopted)
SWID Tags	ISO/IEC 19770 standard

SBOM generation is becoming mandatory for government and regulated industries.

5.3 Real-World Dependency Attacks

Modern applications rely heavily on open-source and third-party libraries, often importing hundreds of packages with just a few lines of code. While this accelerates development, it also introduces hidden risks, especially when these dependencies are not carefully vetted or monitored. Attackers are increasingly targeting this software supply chain by:

- Infiltrating popular packages with malicious code

- Creating look-alike (typosquatted) packages to trick developers

- Compromising build systems to inject harmful dependencies

- Exploiting outdated or vulnerable libraries still in use

These attacks can be silent and devastating, granting attackers access to production environments, sensitive data, or downstream users. Understanding real-world incidents helps illustrate the urgency of dependency management and the need for tools like Software Composition Analysis (SCA).

The table below outlines major dependency attacks that made headlines, including their impact, method, and what went wrong:

Incident	Attack Summary
Log4Shell (2021)	Zero-day in ubiquitous Log4j Java library — impacted thousands of orgs
SolarWinds (2020)	Malicious code injected into Orion build — supply chain compromise
event-stream (2018)	Malicious npm package update exposed private keys
LeftPad (2016)	Removal of one tiny npm package broke thousands of projects

Your weakest dependency may not even be in your direct code.

5.4 Where SCA Fits in CI/CD Pipelines

Software Composition Analysis (SCA) tools are essential in modern Continuous Integration/Continuous Deployment (CI/CD) pipelines, helping teams automatically detect known vulnerabilities in third-party libraries and dependencies. These tools analyze application manifests (e.g., package.json, requirements.txt, pom.xml) to identify which open-source components are in use and whether any of them pose a risk due to outdated versions, licensing issues, or reported CVEs (Common Vulnerabilities and Exposures).

To be effective, SCA should be integrated at multiple stages of the DevOps workflow:

- At commit time, to block risky libraries early

- During builds to scan the complete dependency tree

- Post-deployment to monitor for newly discovered vulnerabilities in production

The table below outlines optimal integration points for SCA tools within the CI/CD pipeline and the types of insights they provide at each stage.

SCA must run at multiple stages:

Stage	Automation Integration
Developer IDE	IDE plugins warn of vulnerable packages
Pull Requests	SCA runs on package files (e.g. package.json, requirements.txt)
Build Stage	SBOM generated, new packages scanned
Pre-Deployment	Policy gates block high CVSS vulnerabilities
Post-Deployment	Continuous SCA monitoring for new CVEs

5.5 SCA Tools Overview

To effectively manage risks from third-party and open-source components, organizations rely on Software Composition Analysis (SCA) tools. These tools automatically detect and catalog all dependencies used within a codebase, assess them against known vulnerabilities, and provide actionable insights for remediation. By integrating SCA tools into the CI/CD pipeline, teams gain continuous visibility into the security posture of their software supply chain. The table below outlines some of the most widely adopted SCA tools and their key features:

Tool	Supported Languages	Strengths
Snyk	Multi-language, SaaS & self-hosted	Excellent developer integration
OWASP Dependency-Check	Java, Node.js, .NET, Python	Open-source baseline scanner
WhiteSource (Mend)	Enterprise SaaS platform	Deep licensing + vulnerability management
Anchore	Container-focused SCA	Integrates SBOM, container scanning
Syft/Grype	Lightweight, open-source	SBOM generation + container CVE scanning

Example: OWASP Dependency-Check Integration

Command-line Usage:

bash

CopyEdit

dependency-check.sh --project MyApp --scan /path/to/project

Integrate this into the CI build job before artifact packaging.

5.6 Container Security

As containers become a foundational element of modern DevOps and microservices architectures, securing them is critical to maintaining overall application integrity. Container security focuses on safeguarding the containerized environment across the entire lifecycle, from image creation and scanning to runtime protection and compliance monitoring. This includes identifying vulnerabilities in base images, enforcing least privilege policies, and detecting anomalous behavior at runtime. The following sections explore essential tools, best practices, and integration strategies for achieving robust container security within a SecDevOps pipeline. Container Security involves scanning Docker images and container registries for:

- Known CVEs in packaged OS layers.

- Insecure default configurations.

- Outdated base images.

- Secret leakage inside image layers.

Container vulnerabilities are not just app vulnerabilities — they're full OS & platform risks.

5.7 Container Scanning Tools

Container scanning tools are essential for detecting vulnerabilities, misconfigurations, and outdated packages within container images before they are deployed into production environments. These tools analyze the contents of images, including base layers, application dependencies, and embedded secrets, to ensure compliance with security standards and policies. By integrating container scanning into the CI/CD pipeline, organizations can shift security left and reduce the risk of deploying containers that are vulnerable to exploitation. Below is a list of widely used container scanning tools and their key capabilities:

Tool	Focus Area	Usage
Trivy	Containers & IaC scanning	Fast, easy-to-integrate, open-source
Grype	Containers + SBOM generation	Open-source, CycloneDX SBOM output
Aqua Security (Enterprise)	Full container lifecycle security	Advanced commercial platform

Tool	Focus Area	Usage
Anchore	Open-source & enterprise container security	Deep registry integration

Example: Trivy in CI Pipeline

bash

CopyEdit

trivy image myapp:latest

- Can scan images during the build stage.

- Produces CVSS-scored vulnerability reports.

- Supports GitHub Actions, Jenkins, and GitLab CI.

5.8 Policy Gates for Dependency & Container Security

To enforce security standards and ensure only compliant builds progress through the pipeline, policy gates are established at key stages of the CI/CD process. These gates serve as automated checkpoints that validate open-source dependencies and container images against predefined criteria, including severity thresholds, license policies, and compliance requirements. By implementing policy gates, teams can prevent high-risk components from being deployed, reduce technical debt, and maintain a consistent security baseline. The following table outlines examples of standard policy gate rules and where they are typically applied in the pipeline:

Recommended Security Gates:

Gate Type	Blocking Condition
SCA Gate	CVSS $\geq$ 7.0
Container Gate	Critical or High CVEs in image
SBOM Validation	Missing SBOM file or signature
Secret Gate	Credentials detected in image layers

Security gates enforce non-negotiable quality standards.

5.9 Handling Transitive Dependency Risks

Transitive dependencies, those indirectly included through primary dependencies, pose significant security challenges, as they often introduce vulnerabilities outside the developer's immediate awareness. These hidden components can account for the majority of a software's dependency tree, making it critical to assess and monitor them rigorously. Effective handling of transitive risks involves using advanced SCA tools that recursively analyze all dependency layers, applying strict version controls, and incorporating automated alerts for newly disclosed vulnerabilities. This section examines strategies for gaining visibility into transitive dependencies and proactively mitigating their associated risks. Transitive dependencies are packages indirectly imported by direct packages.

- Often comprise the majority of the dependency tree.

- Harder to monitor manually.

- Require automated tools with deep dependency resolution capabilities.

SCA tools must support full dependency graph traversal to catch deep, hidden vulnerabilities.

5.10 Dependency Security Metrics

Measuring the effectiveness of dependency security efforts is crucial for continuous improvement and risk management. Dependency security metrics offer insight into the health of the software supply chain by tracking key factors, including the number of known vulnerabilities, average time to remediation, transitive risk exposure, and policy violations. These metrics enable teams to identify trends, prioritize remediation efforts, and demonstrate compliance with both internal and external security standards. The following are key metrics that organizations should monitor to assess and improve their dependency and container security posture:

Metric	Interpretation
% of projects scanned	Pipeline SCA coverage
Time to patch critical CVEs	Responsiveness to supply chain risk
Mean time to generate SBOM	SBOM process maturity

Metric	Interpretation
False positive suppression rate	Scanner tuning quality
Number of stale dependencies	Hygiene of update practices

5.11 Final Takeaway

You don't just ship your code, you ship your entire supply chain.

SecDevOps demands visibility into every dependency, third-party library, and containerized workload. Through:

- Automated SCA scanning,

- Continuous SBOM generation,

- Hardened container images,

- Strict pipeline gates,

Organizations can confidently manage the modern software supply chain, proactively protecting both themselves and their customers.

Chapter 6 — Infrastructure as Code (IaC) Security

"In modern environments, your infrastructure is code, and code can be hacked."

6.1 Infrastructure as Code (IaC)

Infrastructure as Code (IaC) is the practice of managing and provisioning computing infrastructure through machine-readable configuration files, rather than manual processes or interactive configuration tools.

- Cloud environments (AWS, Azure, GCP) can be fully defined and deployed using code.

- Infrastructure resources (servers, databases, networks, permissions, firewalls) are declared as version-controlled code files.

- IaC allows automated, consistent, and repeatable deployments.

IaC brings DevOps principles, automation, versioning, and repeatability to infrastructure itself.

Common IaC Languages & Tools

Infrastructure as Code (IaC) enables teams to define, provision, and manage infrastructure through code, resulting in consistent, repeatable, and automated deployments. A variety of languages and tools have emerged to support Infrastructure as Code (IaC) across different platforms, each offering unique strengths in terms of scalability, modularity, and integration. Understanding these tools is essential for selecting the right fit for the environment and embedding security from the ground up. The table below provides an overview of widely used IaC languages and tools, highlighting their key features and typical use cases:

Tool	Use Case
Terraform	Multi-cloud infrastructure provisioning
AWS CloudFormation	AWS-native IaC
Azure ARM / Bicep	Azure-native IaC
Google Deployment Manager	GCP-native IaC
Ansible	Configuration management, provisioning
Pulumi	Code-based IaC (Python, Go, TypeScript)

6.2 Why IaC Creates New Security Risks

While Infrastructure as Code (IaC) enhances speed, consistency, and scalability in infrastructure management, it also introduces new security risks that traditional approaches did not encounter. Since IaC templates define critical infrastructure components, such as networks, firewalls, and storage, misconfigurations or hardcoded secrets in code can lead to severe vulnerabilities across entire environments. Additionally, version control systems may expose sensitive infrastructure logic to unauthorized access if not properly secured. This section examines the primary reasons why Infrastructure as Code (IaC) can increase the attack surface and why securing these artifacts is crucial in any modern DevSecOps strategy.

While IaC enables powerful automation, it also introduces:

Risk	Description
Misconfigurations	Exposing public buckets, open firewalls, weak IAM permissions
Hardcoded Secrets	API keys or credentials stored in templates
Drift	Differences between declared IaC and actual deployed infrastructure
Uncontrolled Reuse	Reusing insecure IaC modules across projects

Risk	Description
Privilege Escalation	Over-provisioned IAM roles or policies

IaC amplifies both speed and risk — security must scale with it.

6.3 Where IaC Security Fits into Pipelines

Integrating IaC security into CI/CD pipelines ensures that infrastructure definitions are continuously validated for security and compliance before deployment. By embedding automated IaC scanning tools early in the development lifecycle, during code commits, pull requests, or pre-deployment stages, teams can detect misconfigurations, policy violations, and embedded secrets before they reach production. This proactive approach helps enforce security-as-code principles and align infrastructure provisioning with organizational governance standards. The following sections outline how and where Infrastructure as Code (IaC) security checks should be integrated across the pipeline stages.

Stage	Security Action
Pull Requests	Validate IaC changes (lint, policy checks, misconfig detection)
CI Builds	Run IaC vulnerability scanners
Deployments	Apply runtime policy enforcement (OPA, Kyverno)

Stage	Security Action
Post-Deployment	Continuously monitor drift and security posture

IaC scanning must shift left into developer workflows.

6.4 Key IaC Security Tools

To effectively manage and mitigate risks introduced by Infrastructure as Code, a range of security tools has emerged to scan, validate, and enforce best practices across IaC templates. These tools analyze configuration files for common misconfigurations, policy violations, and insecure defaults in platforms like Terraform, AWS CloudFormation, Kubernetes manifests, and more. By integrating these tools into development and deployment workflows, teams can shift left and catch security issues early. The table below provides an overview of leading IaC security tools, along with their core features and supported platforms:

IaC Static Scanners

Tool	Supported Formats	Strength
Checkov	Terraform, CloudFormation, ARM, Kubernetes	Fast, highly adopted open-source tool
Tfsec	Terraform	Terraform-specific deep checks

Tool	Supported Formats	Strength
TFLint	Terraform	Syntax, policy, and security linting
CloudFormation Guard	AWS CloudFormation	Native AWS policy validation
KICS	Multiple IaC formats	Broad language coverage

Most tools integrate directly into CI pipelines.

Example: Checkov in CI

bash

CopyEdit

checkov -d ./terraform

- Scans entire IaC codebase for hundreds of security misconfigurations.

- Supports policy-as-code rules (custom or built-in).

6.5 Policy-as-Code (PaC) for IaC Security

Policy-as-Code (PaC) allows organizations to encode security, compliance, and governance rules into machine-readable policies that enforce safe infrastructure deployments.

- Declarative policies run automatically in CI/CD pipelines.

- Prevents human errors before reaching cloud providers.

PaC Engines

Tool	Usage
Open Policy Agent (OPA)	Universal policy engine for IaC, Kubernetes, APIs
OPA Gatekeeper	Kubernetes-native policy enforcement
Kyverno	Kubernetes policy-as-code
HashiCorp Sentinel	Native to Terraform Enterprise

Policy-as-Code turns security rules into enforceable, testable CI/CD controls.

Example: OPA Rego Rule

rego

CopyEdit

```
deny[msg] {
  input.resource.type == "aws_s3_bucket"
  not input.resource.encryption.enabled
  msg := "S3 bucket must have encryption enabled."
}
```

This rule prevents the deployment of unencrypted S3 buckets.

6.6 Common IaC Security Misconfigurations

Misconfigurations in Infrastructure as Code templates are a leading cause of cloud security breaches. These issues often arise from overly permissive access controls, exposed secrets, lack of encryption, or insecure default settings. Because IaC templates define infrastructure at scale, a single misconfiguration can lead to widespread vulnerabilities across environments. Recognizing these common pitfalls is essential for building secure infrastructure from the ground up. The table below outlines some of the most frequent IaC misconfigurations, their potential impact, and recommended remediation strategies:

Misconfiguration	Risk
Public S3 Buckets	Data exfiltration
Open SSH Ports	Brute force access
Weak IAM Policies	Privilege escalation
Missing Logging	Incident response blind spots
Insecure Security Groups	Exposed attack surface
Hardcoded Secrets	Key compromise
Unencrypted Volumes	Data theft risks

Most cloud breaches trace back to misconfigured resources.

6.7 IaC Drift Detection

Drift occurs when the actual state of infrastructure diverges from the desired state defined in Infrastructure as Code templates, often due to manual changes, ad-hoc fixes, or external systems. Undetected drift can introduce security risks, operational inconsistencies, and compliance violations. Drift detection tools help teams monitor and compare live infrastructure against the declared IaC configurations, enabling early identification and correction of unauthorized or unintended changes. The following table highlights key drift detection tools and approaches commonly used to maintain infrastructure integrity:

Drift: When the deployed infrastructure deviates from the declared IaC definition.

Drift Cause	Security Impact
Manual changes	Unknown attack surface
Hotfixes outside IaC	Compliance violations
Policy updates not applied	Missed new controls

Drift Detection Tools

Tool	Use
Terraform Plan	Shows proposed changes before apply
DriftCTL	Open-source drift detection
AWS Config	AWS-native drift monitoring
Azure Resource Graph	Azure drift visibility

IaC only works when you trust that your code reflects reality.

6.8 Continuous Compliance with IaC Security

Achieving and maintaining compliance in cloud-native environments requires more than one-time audits; it demands continuous validation of infrastructure against security policies and regulatory standards. By embedding compliance checks directly into the IaC pipeline, organizations can ensure every deployment adheres to governance rules such as CIS Benchmarks, NIST, or organizational-specific frameworks. This proactive approach helps reduce audit fatigue, detect drift from compliance baselines, and enforce security consistently at scale. Below are key practices and tools that support continuous compliance in IaC workflows:

IaC scanning supports:

- PCI-DSS

- SOC 2

- ISO 27001

- NIST 800-53

- CIS Benchmarks

IaC security allows you to prove compliance continuously, not just at audit time.

6.9 Securing Shared IaC Modules

Shared IaC modules promote reusability, consistency, and efficiency across teams and projects, but they also present unique security risks if not properly managed. Since these modules often serve as foundational building blocks, any vulnerability or misconfiguration within them can propagate across multiple environments. Securing shared modules involves enforcing version control, validating inputs, restricting sensitive outputs, and applying rigorous code reviews. This section outlines best practices and strategies to ensure shared IaC components remain secure, maintainable, and trustworthy. Many teams utilize shared Infrastructure as Code (IaC) modules for reuse across projects. If insecure modules propagate across environments, they become central points of failure.

Best Practices:

- Version control all modules.

- Apply static scanning to module repositories.

- Perform peer code reviews on shared modules.

- Restrict publishing permissions.

Shared IaC modules = shared risk.

6.10 IaC Security Metrics

Measuring the effectiveness of Infrastructure as Code (IaC) security practices is essential for continuous improvement, accountability, and risk reduction. IaC security metrics help teams track progress, identify weak points, and demonstrate compliance with internal policies and external standards. These metrics may include the number of misconfigurations detected, remediation time, policy violation rates, and coverage of IaC scanning tools within the pipeline. The following are key metrics that organizations should monitor to evaluate and enhance the security posture of their IaC practices:

Metric	Meaning
% of PRs scanned for IaC issues	Pipeline coverage
Mean time to fix IaC findings	Responsiveness
# of public resource exposures prevented	Proactive risk reduction
Compliance coverage gap (controls unmet)	Audit readiness maturity
Policy-as-code rule coverage	Governance enforcement scope

6.11 Final Takeaway

In cloud-native security, misconfiguration is the #1 threat.

IaC security allows teams to:

- Catch risks before deployment.

- Enforce security automatically through pipelines.

- Codify compliance into version-controlled rules.

- Scale safely across multi-cloud environments.

In SecDevOps, infrastructure is code, and code must be secured.

Chapter 7 — Secrets Management and Identity Security

"In cloud-native systems, your biggest vulnerability often isn't code — it's credentials."

7.1 Introduction: Why Secrets Are a Top Attack Vector

Secrets, such as API keys, database credentials, SSH keys, encryption keys, and cloud tokens, are often the crown jewels of the environment.

- If secrets are leaked, attackers gain full access to critical systems.

- Public GitHub repos, logs, container layers, and CI/CD pipelines often accidentally expose credentials.

- Breaches like Uber (2022), CodeSpaces, and Slack (2022) demonstrate the catastrophic impact of leaked secrets.

In SecDevOps, secrets must be treated as first-class security objects, not as environment variables or config files.

7.2 Secrets Management

Secrets management refers to securely storing, distributing, rotating, auditing, and revoking credentials, keys, and sensitive configuration data throughout the entire Software Development Life Cycle (SDLC).

No hardcoded credentials, no untracked access, complete lifecycle control.

7.3 Where Secrets Live Across Your Systems

Secrets—such as API keys, passwords, tokens, and encryption keys—can be found across many layers of modern infrastructure and application stacks. Identifying where these secrets reside is the first step in building a comprehensive secrets management strategy. From configuration files and environment variables to container images and CI/CD pipelines, secrets are often scattered and may be inadvertently exposed. Understanding their distribution helps teams implement targeted security controls, reduce attack surfaces, and avoid accidental leaks. The table below outlines common locations where secrets are typically found in software systems:

Layer	Common Secrets
Application Code	API keys, DB passwords
IaC Templates	AWS Access Keys, SSH keys
CI/CD Pipelines	Deployment tokens, artifact signing keys
Containers	Embedded config secrets
Runtime	Cloud credentials, session tokens
Developer Machines	SSH keys, personal access tokens

Secrets sprawl creates one of the largest blind spots in cloud-native security.

7.4 Secrets Leakage Risks

Leaked secrets can lead to severe security breaches, including unauthorized access to critical systems, data exfiltration, and service disruption. Secrets often end up exposed through source code repositories, logs, container images, or misconfigured CI/CD pipelines. When not properly managed, even a single leaked credential can serve as an entry point for attackers. This section outlines the most common scenarios where secrets leakage occurs and the potential impact on system integrity and organizational security. The table below summarizes key leakage vectors and associated risks:

Leakage Vector	Example
Public GitHub commits	Developers accidentally commit .env files
Docker images	Secrets baked into image layers
Build logs	Verbose logs printing keys
CI/CD variables	Insecure storage or unrestricted access
Third-party tools	SaaS integrations mishandling keys

7.5 Secure Secrets Storage Principles

Properly storing secrets is fundamental to preventing unauthorized access and maintaining the confidentiality, integrity, and availability of sensitive systems. Secure secrets storage goes beyond simple encryption—it requires a structured approach that includes access controls, audit logging, segregation of duties, and automated key rotation. By adhering to well-defined principles, organizations can ensure that secrets are protected throughout their lifecycle, whether at rest, in transit, or during use. Below are the core principles that should guide the secure storage and management of secrets:

Core Principles:

- Centralized Vaulting — No credentials stored directly in code or config files.
- Role-Based Access Control (RBAC) — Least privilege access to secrets.
- Dynamic Secrets — Generate short-lived credentials on demand.
- Rotation — Regularly rotate and expire secrets automatically.
- Auditability — Full logging of who accessed which secrets and when.

Secrets are dynamic assets that require lifecycle management, not static configuration.

7.6 Secrets Vaulting Solutions

Secrets vaulting solutions are purpose-built tools designed to securely store, access, and manage sensitive credentials such as API keys, passwords, certificates, and tokens. These tools offer centralized control, encryption at rest and in transit, access policies, audit logging, and automated secret rotation. Integrating vaulting solutions into your development and deployment workflows reduces the risk of secrets sprawl and helps enforce security best practices. The table below compares popular secrets vaulting solutions, highlighting their key features, integrations, and use cases:

Tool	Type	Strengths
HashiCorp Vault	Self-hosted, enterprise vault	Dynamic secrets, PKI, highly scalable
AWS Secrets Manager	AWS-native	Easy integration, automatic rotation
AWS Parameter Store	AWS-native	Lightweight key/value secrets store
Azure Key Vault	Azure-native	Strong RBAC integration
GCP Secret Manager	GCP-native	IAM-based fine-grained controls

Tool	Type	Strengths
Doppler, Akeyless, 1Password Secrets	SaaS secrets managers	Multi-cloud flexibility, team collaboration

Example: Fetching a Secret from AWS Secrets Manager (Python)

python

CopyEdit

import boto3

client = boto3.client('secretsmanager')

response = client.get_secret_value(SecretId='MyDatabaseSecret')

credentials = response['SecretString']

Applications fetch secrets dynamically at runtime rather than storing them locally.

7.7 Integrating Secrets Management into CI/CD

Integrating secrets management into CI/CD pipelines is critical for automating secure access to credentials without exposing them in code or logs. By embedding secrets retrieval and injection processes into pipeline stages, teams can ensure that sensitive information is handled securely throughout the build, test, and deployment phases. This integration typically involves using secure environment variables, external vaults, and access controls to prevent accidental leaks or misuse. The table below outlines common integration patterns and best practices for securely managing secrets within CI/CD workflows:

Stage	Practice
CI Build	Use secrets injection plugins (e.g., Vault Agent)
Deployment	Fetch secrets at runtime, not hardcoded in artifacts
Pipelines	Access tokens scoped per pipeline/job
Secrets Rotation	Rotate keys automatically upon each deployment

CI/CD pipelines are high-risk zones for secrets leakage — they require extra scrutiny.

7.8 Secrets Scanning for Prevention

To proactively prevent secrets exposure, organizations must implement automated secrets scanning as part of their development and deployment workflows. Secrets scanners analyze source code, configuration files, and version control history to detect hardcoded credentials, tokens, and other sensitive data before they are committed or deployed. Early detection helps reduce the risk of accidental leaks and enforces secure coding practices across teams. The table below presents common secrets scanning tools and their capabilities for preventing credential exposure:

Secrets Scanning Tools

Tool	Usage
Gitleaks	Git repository secret scanning
TruffleHog	High-entropy string scanning
GitGuardian	SaaS platform for Git + CI pipelines
Talisman	Local pre-commit hook scanning

Preventing secrets from entering your repos is always easier than revoking compromised ones.

7.9 Identity Security

Identity security protects credentials, accounts, roles, and permissions, ensuring the right entity has the right access to the right resource at the right time.

- Strong identity controls enforce the principle of least privilege (PoLP).

- Identity compromise often leads to lateral movement and privilege escalation.

In cloud-native systems, identity is the new perimeter.

7.10 IAM (Identity and Access Management) Design

A well-designed Identity and Access Management (IAM) framework is essential for controlling who has access to what resources—and under what conditions—within modern DevSecOps environments. IAM design involves defining roles, permissions, authentication mechanisms, and policies that align with the principle of least privilege and enforce strong access boundaries. A robust IAM strategy reduces the risk of unauthorized access, insider threats, and privilege escalation. The following key principles outline best practices for designing secure and scalable IAM architectures:

Key IAM Security Principles:

Principle	Description
Least Privilege	Grant only required permissions
Separation of Duties	Split access across roles (e.g. read-only vs. write access)
Role-Based Access Control (RBAC)	Group permissions into manageable roles
Just-In-Time Access	Temporary elevation rather than permanent privileges
Credential Expiration	Use time-limited credentials
MFA Everywhere	Multi-factor authentication for all administrative accounts

IAM design flaws often result in catastrophic full-environment compromises.

Example: AWS IAM Policy for Least Privilege

json

CopyEdit

```
{
    "Effect": "Allow",
    "Action": ["s3:GetObject"],
    "Resource": ["arn:aws:s3:::my-secure-bucket/*"]
}
```

Avoid wildcard **"" permissions whenever possible.***

7.11 Cloud Identity Risks

As organizations adopt cloud platforms, identity becomes a critical security perimeter. Misconfigured identities, overly permissive roles, and a lack of visibility into access behaviors can lead to serious security incidents, including unauthorized access, data exfiltration, and privilege escalation. Cloud identity risks are often amplified by the dynamic nature of cloud environments and the complexity of managing identities at scale. The table below outlines common cloud identity risks, their potential impacts, and examples of how these risks can manifest in real-world environments:

Cloud Identity Risk	Potential Impact	Real-World Example
Overly Permissive IAM Roles	Excessive access increases risk of data exfiltration, privilege escalation, or abuse	A developer assigned full admin rights accidentally deletes production resources
Lack of Role Segregation	Blurs responsibilities and reduces accountability	One user with dev/test/prod access pushes untested code directly into production
Misconfigured Federated Identities	Unauthorized access via weak identity provider settings	A misconfigured SAML integration allows external users to log in with elevated rights
Inactive or Orphaned Accounts	Compromised old accounts can provide silent access to attackers	A former employee's cloud account remains active and is later used in a data breach
Hardcoded Cloud Access Keys	Keys exposed in code can lead to unauthorized access	AWS keys committed to GitHub are

Cloud Identity Risk	Potential Impact	Real-World Example
		harvested and used for crypto mining
Lack of MFA Enforcement	Increases susceptibility to credential stuffing and phishing attacks	An attacker logs in using leaked credentials due to missing MFA on a privileged account
Excessive Third-Party Access	Vendors or services gain broad access to internal systems	A third-party CI/CD tool with admin access is compromised, impacting production assets
Unmonitored Temporary Credentials	Temporary credentials can be misused if not tracked or expired properly	A temporary token issued for debugging is reused by an attacker after hours

7.12 Zero Trust Identity Architectures

Zero Trust Identity Architectures operate on the principle that no user or system, whether inside or outside the network perimeter, should be inherently trusted. Instead, identity must be continuously verified through context-aware, risk-based access controls. In a Zero Trust model, access decisions are enforced dynamically based on user identity, device posture, location, behavior, and other contextual signals. This approach significantly reduces the risk of lateral movement and unauthorized access in cloud-native and hybrid environments.

Never assume trust by network location; verify identity for every request.

Zero Trust Identity Pillars:

- Continuous authentication

- Identity federation

- Context-aware access (device, location, risk)

- MFA enforcement

- Strong session expiration and revocation

Example Technologies:

- AWS Cognito

- Azure AD Conditional Access

- Okta Identity Engine

- Google BeyondCorp

7.13 Secrets & Identity Security Metrics

Tracking meaningful metrics is essential for evaluating the effectiveness of secrets management and identity security practices. These metrics provide visibility into how well secrets are protected, how access is managed, and where potential gaps may exist. By continuously monitoring these indicators, security teams can identify trends, enforce compliance, and drive improvements in both operational efficiency and security posture. The table below presents key metrics organizations should monitor to assess their secrets and identity management programs:

Metric	What It Measures
Secrets committed to repos	Hygiene failure points
Time to revoke exposed secrets	IR response time
Secrets rotation frequency	Credential management maturity
% of accounts with MFA enforced	Identity hardening discipline
Number of over-permissioned IAM roles	Least privilege enforcement gaps

7.14 Final Takeaway

Credentials don't fail open — humans do.

By automating:

- Secrets storage and rotation,
- IAM policy enforcement,
- Secrets scanning in pipelines,

- Continuous identity monitoring,

… SecDevOps teams dramatically shrink the attack surface, even as software scales across complex, distributed, multi-cloud environments.

Part III: Pipeline Security and Continuous Compliance

This section focuses on protecting the CI/CD pipeline itself and establishing continuous compliance through policy-as-code and automation.

- Chapter 8: Securing CI/CD Pipelines
  Discusses build runner security, policy gates, artifact integrity, and threat modeling.

- Chapter 9: Compliance as Code and Audit Readiness
  Introduces Compliance as Code, continuous validation, generating audit evidence, and regulatory mapping (e.g., SOC 2, ISO 27001).

Chapter 8 — Securing CI/CD Pipelines

"If an attacker owns your pipeline, they own your entire production environment."

8.1 Introduction: Why CI/CD Pipelines Are High-Value Targets

CI/CD pipelines are the backbone of modern software delivery, enabling rapid and automated code integration, testing, and deployment. However, this speed and automation also make them prime targets for attackers. These pipelines often have access to sensitive assets—such as source code, secrets, cloud credentials, and production environments—making them a powerful vector for supply chain attacks if compromised. A breach in the CI/CD pipeline can allow attackers to inject malicious code, exfiltrate intellectual property, or gain persistent access to critical infrastructure. As such, securing CI/CD pipelines is not optional—it is a foundational requirement for any robust DevSecOps program.

Continuous Integration and Continuous Deployment (CI/CD) pipelines are at the heart of modern software delivery:

- They build, test, and deploy code automatically.

- They hold privileged credentials, signing keys, and deployment secrets.

- They have deep access to both codebases and production infrastructure.

If CI/CD pipelines are compromised, attackers can inject malicious code, poison builds, or directly impact production systems.

8.2 CI/CD Security

CI/CD Security refers to protecting the full software build, test, and deployment automation chain, ensuring:

- Only trusted code is built.

- Only authorized personnel control deployments.

- Secrets and credentials are secured.

- Artifacts are validated before release.

- Pipelines are hardened against insider or external abuse.

The pipeline itself becomes part of your security perimeter.

8.3 The CI/CD Attack Surface

The CI/CD pipeline consists of numerous interconnected tools, services, and scripts—each representing a potential entry point for attackers. From source code repositories and build servers to deployment scripts and artifact registries, every component in the pipeline can be exploited if not properly secured. Understanding the full breadth of the CI/CD attack surface is critical for implementing effective security controls and minimizing exposure. The table below outlines common attack vectors within CI/CD environments and provides descriptions of how adversaries exploit them:

Attack Vector	Description
Compromised SCM (Git)	Malicious commits, poisoned source code
Compromised Build Servers	Pipeline manipulation, backdoor insertion
Secrets Leakage	Credentials exposed in pipeline logs or configs
Third-Party Integrations	OAuth apps with excessive permissions
Supply Chain Poisoning	Compromised dependencies or build plugins
Runner Abuse	Public runners used for crypto mining or lateral movement

CI/CD security must protect both the pipeline infrastructure AND the code flowing through it.

8.4 Where CI/CD Security Fits in SecDevOps

In a SecDevOps framework, security must be tightly integrated into every phase of the CI/CD pipeline—from code commit to deployment. Rather than being an afterthought, security becomes a shared responsibility across development, operations, and security teams. Embedding security controls into the CI/CD workflow ensures that vulnerabilities, misconfigurations, and policy violations are detected and addressed in real time, without slowing down delivery. The table below maps various stages of the CI/CD pipeline to recommended security controls, highlighting where and how security should be enforced:

CI/CD Stage	Security Control
Source Control	PR approvals, signed commits
Build Stage	SAST, secrets scanning, SBOM generation
Artifact Signing	Cryptographic integrity validation
Deployment	IaC scanning, runtime policy validation
Post-Deployment	Monitoring, rollback capability

Security gates integrate into every phase of the delivery process.

8.5 Securing CI/CD Platforms

CI/CD platforms, such as Jenkins, GitLab CI/CD, GitHub Actions, and CircleCI, are central to modern software delivery workflows; as such, they must be hardened against misuse and compromise. These platforms often operate with elevated privileges, integrate with multiple systems, and automate sensitive tasks, making them attractive targets for attackers. Securing these platforms involves applying access controls, patch management, audit logging, and plugin vetting, among other best practices. The table below highlights popular CI/CD platforms along with common security considerations and recommended hardening measures:

CI/CD Platforms

Platform	Examples
Cloud-native	GitHub Actions, GitLab CI, Bitbucket Pipelines
SaaS	CircleCI, TravisCI, Jenkins Cloud
Self-hosted	Jenkins, Bamboo, TeamCity, Spinnaker

Core Platform Hardening Steps

To effectively defend CI/CD platforms against exploitation, it's essential to implement a set of core hardening controls that address both configuration weaknesses and operational risks. These steps help reduce the attack surface, enforce least privilege, and ensure consistent enforcement of security policies. From restricting administrative access to isolating build environments, each control contributes to building a resilient CI/CD foundation. The table below outlines key hardening controls along with their purpose in securing the CI/CD platform:

Control	Purpose
Enforce MFA	Strong authentication for all users
Least Privilege RBAC	Minimize who can modify pipelines
Secrets Vault Integration	No hardcoded secrets
Audit Logging	Visibility into pipeline changes
IP Restrictions	Limit access to CI/CD servers
Dependency Isolation	Prevent build dependencies from accessing internet unnecessarily

CI/CD systems must be treated as production-critical security assets.

8.6 Hardening Build Runners

Build runners are the execution engines of CI/CD pipelines, responsible for compiling code, running tests, and deploying artifacts. Because they handle untrusted code and often have access to secrets and production credentials, securing them is critical to preventing pipeline compromise. Hardening build runners involves isolating workloads, restricting network access, limiting privileges, and ensuring a clean and reproducible environment for each job. The table below outlines essential hardening controls for build runners and their corresponding security benefits:

125

Hardening Control	Description
Ephemeral Runners	Use one-time VMs/containers for each job
Network Isolation	Prevent direct internet access unless required
Resource Scoping	Limit file system access to workspace only
Non-Root Execution	Avoid privileged execution contexts
Containerized Builds	Run builds inside isolated Docker containers

Never allow long-lived, shared runners for sensitive builds.

8.7 Signing Build Artifacts (Software Supply Chain Integrity)

As software supply chain attacks become increasingly sophisticated, ensuring the authenticity and integrity of build artifacts is vital. Artifact signing provides cryptographic proof that software components originate from a trusted source and have not been tampered with during the build or delivery process. This protects downstream users, enforces accountability, and helps comply with regulatory and industry standards. The list below outlines the key reasons why artifact signing is a critical component of CI/CD security:

Why Artifact Signing Matters:

- Ensures build outputs haven't been tampered with.

- Verifies build provenance (who built it, with what code).

- Prevents supply chain injection attacks.

Artifact Signing Tools

Tool	Use Case
Sigstore (Cosign, Rekor)	Modern keyless signing (widely adopted)
The Update Framework (TUF)	Secure update delivery frameworks
Notary (Docker Content Trust)	Docker image signing

Example: Signing a Docker Image with Cosign

bash

CopyEdit

*cosign sign --key cosign.key docker.io/myorg/myapp:1.0.*0

Signed artifacts enable downstream deployment verification.

8.8 Preventing CI/CD Pipeline Abuse

I/CD pipelines, if improperly secured, can be exploited to perform unauthorized actions such as cryptomining, data exfiltration, malware injection, or privilege escalation. Attackers may abuse misconfigured jobs, compromised credentials, or overly permissive scripts to turn the pipeline into an attack vector. Proactively identifying and mitigating these abuse scenarios is essential to maintain the integrity of the software delivery process. The table below highlights common types of CI/CD pipeline abuse and corresponding mitigation strategies:

Abuse Type	Mitigation
Crypto Mining	Job runtime limits, resource quotas
Code Injection	Signed commits, PR approvals
Secret Exfiltration	Strict log redaction, encrypted secrets
Lateral Movement	Network isolation of runners
Malicious PR Automation	Require multiple reviewers for sensitive repos

8.9 Security Gates in CI/CD Pipelines

Security gates act as automated enforcement points within CI/CD pipelines, designed to halt or fail a build when defined security conditions are not met. These gates ensure that critical security checks—such as vulnerability scans,

policy compliance, license validation, and test coverage—
are treated as mandatory steps rather than optional ones. By
integrating security gates early and consistently, teams can
prevent insecure code or configurations from progressing
through the pipeline. The table below outlines common
types of security gates and the blocking conditions that
trigger them:

Gate Type	Blocking Condition
PR Checks	Signed commits, passing SAST
Build Gates	No critical SCA or IaC misconfigs
Secrets Scan	No committed credentials
SBOM Validation	No banned or vulnerable packages
Policy Gates	Compliance rules enforced pre-deployment
Deployment Gates	DAST pass rate for staging environments

Security gates should be codified and version-controlled as part of the pipeline definition.

8.10 CI/CD Security Metrics

To effectively manage and improve the security of CI/CD pipelines, organizations must track meaningful metrics that reflect the health, coverage, and responsiveness of their security practices. These metrics provide visibility into pipeline risks, measure the success of implemented controls, and help prioritize remediation efforts. Monitoring the right indicators allows teams to detect trends, demonstrate compliance, and continuously optimize their DevSecOps posture. The table below lists key CI/CD security metrics and what each one measures:

Metric	What It Measures
% of builds scanned	Pipeline security coverage
% of artifacts signed	Supply chain protection strength
Secret leakage incidents	Hygiene discipline
MTTR for CI/CD vulnerabilities	Response readiness
Number of blocked malicious PRs	Preventative strength
Rollback success rates	Deployment resilience

8.11 CI/CD Threat Modeling Checklist

Threat modeling helps identify, understand, and mitigate potential security threats across the CI/CD pipeline before they can be exploited. By systematically analyzing each component and phase of the pipeline, teams can uncover weak points, assess attack vectors, and apply appropriate countermeasures. A CI/CD-specific threat modeling checklist ensures that no part of the delivery process is overlooked and that security is addressed in a proactive and structured manner. The table below maps critical areas of the pipeline to key threats that should be modeled and mitigated:

Pipeline Area	Key Threats to Model
SCM	Commit spoofing, PR poisoning
Build Agents	Credential theft, runner compromise
Artifact Stores	Integrity loss, registry poisoning
Deployment	Unauthorized environment access
Secrets Storage	Exposure in logs, backups
Third-Party Integrations	OAuth abuse, plugin vulnerabilities

Every CI/CD component becomes an attack surface that must be modeled.

8.12 Final Takeaway

Your pipelines are no longer simple automation scripts —
they are part of your core production environment.

In SecDevOps:

- CI/CD security is not optional.

- Pipelines must enforce security gates at every stage.

- Supply chain protection begins at build time, not release time.

- Hardened, monitored, and signed pipelines give you resilience at scale.

Chapter 9 — Compliance as Code and Audit Readiness

"In SecDevOps, compliance is no longer a documentation exercise — it's a continuous, automated control system."

9.1 Introduction: Why Compliance Must Evolve

Traditional compliance models were designed for static, perimeter-based environments—far removed from today's dynamic, cloud-native, and continuously deployed systems. As organizations embrace DevSecOps, infrastructure as code, and automated delivery pipelines, compliance must evolve from point-in-time audits to continuous, code-integrated enforcement. Static checklists and manual reviews can no longer keep pace with rapid change. Instead, modern compliance requires real-time monitoring, policy-as-code, and seamless integration with development workflows. This section explores why compliance strategies must adapt to support speed, scalability, and security in modern software delivery.

Traditional compliance models struggle to keep up with modern cloud-native, microservice-based, fast-release software environments.

- Compliance teams often rely on manual checklists, spreadsheets, and periodic audits.

- DevOps teams push code daily or hourly, creating a moving target for auditors.

- Regulatory frameworks (SOC 2, ISO 27001, PCI-DSS, GDPR, HIPAA, NIST) demand consistent evidence of security controls.

SecDevOps bridges this gap by shifting compliance controls into automated pipelines, codified as machine-enforceable rules.

9.2 Compliance as Code

Compliance as Code (CaC) is the practice of embedding regulatory and security control requirements directly into your infrastructure, code, pipelines, and monitoring systems — making compliance continuous, automated, and provable.

Every commit, build, and deployment automatically enforces compliance policies.

9.3 Why Compliance as Code Is Critical

In fast-paced DevSecOps environments, manual compliance processes are no longer sufficient to ensure consistent enforcement and auditability. Compliance as Code transforms static policies into machine-readable, testable configurations that can be integrated directly into CI/CD pipelines and infrastructure provisioning workflows. This approach enables real-time policy validation, automated enforcement, and continuous monitoring, eliminating delays and reducing human error. By codifying compliance, organizations can align security and governance with development velocity. The table below contrasts traditional compliance approaches with the modern Compliance as Code paradigm:

Traditional Compliance	Compliance as Code
Periodic audits	Continuous, real-time compliance
Manual checklists	Automated controls
Reactive findings	Proactive prevention
Spreadsheet evidence	Immutable audit trails
High cost of audits	Reduced audit burden

9.4 Key Regulatory Frameworks Covered by SecDevOps

Modern software systems often operate under strict regulatory requirements that govern data protection, access control, operational resilience, and more. SecDevOps practices help bridge the gap between rapid development and regulatory compliance by embedding security and governance controls into automated workflows. By aligning with key regulatory frameworks, organizations can reduce risk, avoid penalties, and ensure trust with customers and stakeholders. The table below highlights major regulatory standards and their primary focus areas relevant to SecDevOps implementations:

Standard	Focus Area
SOC 2	SaaS security & availability
ISO 27001	Global information security standards
PCI-DSS	Payment card processing
HIPAA	Healthcare data privacy
GDPR	EU personal data protection
NIST 800-53 / FedRAMP	U.S. federal security controls
CIS Benchmarks	Cloud security configurations

Compliance as Code enables cross-mapping multiple frameworks simultaneously.

9.5 Where Compliance as Code Lives in Pipelines

To be effective, Compliance as Code must be seamlessly integrated at multiple stages of the CI/CD pipeline. From infrastructure provisioning and code analysis to deployment and runtime monitoring, compliance checks should be automated and continuous. This ensures that security and regulatory policies are enforced consistently without slowing down development. By embedding compliance controls early and throughout the pipeline, organizations can detect violations in real time and maintain an auditable trail of enforcement actions. The table below outlines where Compliance as Code typically operates within the pipeline and the types of controls applied at each stage:

Pipeline Stage	Control Type
SCM	PR security reviews, code ownership enforcement
Build	SBOM generation, secrets scanning
Deployment	IaC scanning, policy-as-code enforcement
Runtime	Continuous monitoring, logging, anomaly detection
Incident Response	Automated response runbooks, forensics

9.6 Core Tools for Compliance as Code

Implementing Compliance as Code requires specialized tools that can automatically validate infrastructure and application configurations against predefined policies. These tools support policy-as-code frameworks, integrate with CI/CD pipelines, and enforce compliance across cloud environments, containers, and IaC templates. By codifying compliance rules, these solutions help teams detect violations early, maintain continuous governance, and align with regulatory and organizational standards. The table below presents widely used Compliance as Code tools along with their typical usage and areas of coverage:

IaC Compliance Tools

Tool	Usage
Checkov	Cloud compliance rule checks
tfsec	Terraform compliance scanning
Open Policy Agent (OPA)	Policy-as-Code engine
Regula	Rego-based IaC compliance framework
KICS	Multi-format IaC compliance scanning

Policy-as-Code

Policy-as-Code enables organizations to express security, compliance, and operational rules in declarative, machine-readable formats that can be automatically enforced within CI/CD pipelines and cloud infrastructure. These policies can cover a wide range of controls—from access restrictions and encryption requirements to naming conventions and resource limits. By integrating these policies directly into development workflows, teams can catch violations early and ensure consistent governance. Below are examples of common Policy-as-Code rules used across various platforms and environments:

Policy-as-Code Examples

- Block creation of public S3 buckets.

- Enforce mandatory encryption for databases.

- Prohibit untagged resources (asset inventory control).

Example: OPA Compliance Rule

rego

CopyEdit

```
deny[msg] {
  input.resource.type == "aws_s3_bucket"
  input.resource.public_access == true
  msg := "Public S3 buckets are not allowed."
}
```

OPA converts policies directly into machine-enforced code.

Runtime Compliance & Monitoring Tools

Ensuring compliance doesn't end at deployment—organizations must continuously monitor running systems to detect violations, enforce policies, and maintain audit readiness. Runtime compliance and monitoring tools provide real-time visibility into workloads, configurations, access controls, and behaviors across cloud and containerized environments. These tools help enforce

139

security baselines, detect drift, and generate compliance reports aligned with industry standards. The table below highlights popular runtime compliance and monitoring tools along with their primary use cases and capabilities:

Tool	Usage
AWS Config	Continuous cloud resource compliance
Azure Security Center	Cloud resource governance
Google Security Command Center	GCP compliance visibility
CIS Benchmarks	Cloud provider best practice standards

Cloud-native tools provide real-time drift detection and control enforcement.

9.7 Generating Continuous Evidence for Auditors

In modern DevSecOps environments, auditors require ongoing, verifiable proof that security and compliance controls are being enforced, not just point-in-time snapshots. Generating continuous evidence involves automatically collecting logs, policy evaluations, configuration states, and control results throughout the development and deployment lifecycle. This approach not only streamlines audit readiness but also enhances transparency, accountability, and trust. The table below outlines key types of continuous evidence, their sources, and how they support audit and compliance objectives:

Type of Evidence	Source	Audit/Compliance Objective Supported
Infrastructure Configuration Snapshots	IaC tools (e.g., Terraform, Pulumi), Cloud APIs	Demonstrate secure provisioning and adherence to configuration baselines
Policy Evaluation Logs	Compliance as Code tools (e.g., OPA, Sentinel)	Provide proof of policy enforcement at build and deployment stages
Access Logs & IAM Events	Cloud provider logs (e.g., AWS CloudTrail, Azure AD), IAM tools	Show who accessed what, when, and how—supporting access control reviews
Vulnerability Scan Reports	SAST, DAST, SCA tools (e.g., Snyk, Trivy, SonarQube)	Evidence of proactive threat detection and remediation
Pipeline Audit Trails	CI/CD platforms (e.g., Jenkins, GitLab, GitHub Actions)	Validate that security checks and approvals are integrated and enforced
Runtime Security Alerts	Monitoring tools (e.g., Falco, AWS GuardDuty)	Indicate real-time detection and response to anomalous or policy-violating behavior

Type of Evidence	Source	Audit/Compliance Objective Supported
Drift Detection Logs	Tools like Terraform Drift, AWS Config	Confirm that infrastructure state aligns with declared configurations
Encryption & Data Protection Status	Cloud dashboards, encryption tools, vaults	Validate compliance with data protection regulations (e.g., GDPR, HIPAA)
Change Management Records	Git repositories, issue tracking tools (e.g., Jira)	Demonstrate traceability of code and configuration changes
Automated Compliance Reports	Aggregated reporting tools (e.g., Prisma Cloud, Wiz, Azure Security Center)	Provide auditors with structured, exportable evidence aligned with frameworks

Evidence Type	How CaC Produces It
Access control	IAM policy reviews, logs
Encryption enforcement	IaC rules, build validations
Secure code reviews	PR audit trails

Evidence Type	How CaC Produces It
SBOMs	Build artifact metadata
Incident response testing	Playbook execution logs
Data retention logs	Automatic data lifecycle policies

Evidence is generated automatically and version-controlled — no more scrambling before audits.

9.8 Mapping Controls to Pipelines

To ensure security and compliance are embedded effectively, organizations must map specific controls to the appropriate stages of the CI/CD pipeline. This approach helps operationalize policies by assigning actionable checks—such as access validation, vulnerability scanning, and configuration enforcement—at the right moments in the software delivery lifecycle. By aligning controls with pipeline phases, teams can create a traceable and enforceable security model that integrates seamlessly with development workflows. The table below illustrates how key controls can be mapped to various stages of the pipeline:

Control Type	Pipeline Enforcement
Code Commit	Signed commits, PR approvals
Build Process	SAST, secrets scanning, SBOM generation

Control Type	Pipeline Enforcement
Artifact Security	Artifact signing (Cosign, Sigstore)
Deployment	IaC scanning, policy validation
Runtime	Logging, monitoring, and continuous drift detection

9.9 Example: SOC 2 Control Enforcement in SecDevOps

SOC 2 compliance focuses on the Trust Services Criteria—security, availability, processing integrity, confidentiality, and privacy. In a SecDevOps context, these criteria can be enforced through automated controls and continuous monitoring across the CI/CD pipeline. By mapping SOC 2 requirements to specific technical implementations, organizations can demonstrate ongoing adherence without interrupting delivery speed. The table below provides examples of how key SOC 2 controls can be enforced using SecDevOps practices:

SOC 2 Principle	SecDevOps Control
Logical Access	RBAC enforced via IAM & CI/CD RBAC
Change Management	PR reviews, signed commits, CI logs

SOC 2 Principle	SecDevOps Control
System Monitoring	Cloud SIEM integration
Data Encryption	IaC rules for encryption at rest & transit
Incident Response	Automated playbooks, forensics logging

Regulatory frameworks map naturally to technical controls in automated pipelines.

9.10 Compliance Drift and Continuous Validation

Compliance drift occurs when systems gradually fall out of alignment with established security and regulatory policies, often due to manual changes, misconfigurations, or overlooked updates in dynamic environments. Left unchecked, this drift can lead to audit failures, security exposures, and reputational damage. Continuous validation addresses this risk by automatically monitoring configurations, policies, and controls in real time to detect deviations as they occur. By integrating continuous validation into SecDevOps workflows, organizations can maintain a consistent compliance posture, reduce audit fatigue, and respond swiftly to violations.

Even with Compliance as Code, drift may still occur due to:

Drift Cause	Prevention
Manual changes	Disable console changes; enforce IaC pipelines
Third-party integrations	Review SaaS vendor compliance posture
IAM permission creep	Run scheduled IAM audits
Unexpected service changes	Subscribe to cloud provider change notifications

Automated compliance drift detection closes audit gaps early.

9.11 Audit Readiness Dashboards

Audit readiness dashboards provide real-time visibility into the organization's compliance posture by aggregating key security metrics, control statuses, policy violations, and evidence logs in a centralized interface. These dashboards enable security and compliance teams to proactively identify gaps, track remediation efforts, and demonstrate ongoing adherence to regulatory frameworks. By integrating with CI/CD pipelines, cloud environments, and compliance-as-code tools, audit readiness dashboards help bridge the gap between technical enforcement and executive reporting—streamlining both internal reviews and external audits

Modern audit dashboards consolidate:

- Control status (pass/fail)

- Drift events

- Open findings with owners assigned

- Compliance evidence artifacts

- Audit trails from pipelines and runtime

Example Tools:

Platform	Usage
Prisma Cloud	Multi-cloud compliance
Wiz	Cloud posture visibility
Lacework	Runtime & cloud compliance
AWS Audit Manager	Automated control mapping for AWS

9.12 SecDevOps Compliance Anti-Patterns to Avoid

While integrating compliance into DevSecOps offers significant benefits, there are common pitfalls—or anti-patterns—that can undermine its effectiveness. These often stem from treating compliance as a one-time task, relying solely on manual processes, or introducing controls that hinder development velocity. Recognizing and avoiding these anti-patterns is crucial to maintaining a balanced, scalable, and secure compliance posture. The table below outlines common SecDevOps compliance anti-patterns and explains why they pose risks to both security and operational efficiency:

Anti-Pattern	Why It's Dangerous
Relying solely on manual checklists	Too slow for CI/CD velocity
"Compliance freeze" deployment gates	Blocks releases unnecessarily
Security as separate silo from DevOps	Breaks ownership and speed alignment
Compliance after deployment	Creates blind spots
No evidence automation	Leads to expensive, error-prone audit prep

Compliance succeeds best when built into developer workflows natively.

9.13 Compliance as Code Metrics

To evaluate the effectiveness and maturity of Compliance as Code initiatives, organizations must track specific metrics that reflect policy coverage, enforcement consistency, and overall compliance health. These metrics help quantify progress, uncover gaps, and drive continuous improvement in both security and governance. By integrating metric collection into CI/CD workflows, teams can monitor compliance performance in real time and provide auditors with objective, data-driven evidence. The table below highlights key Compliance as Code metrics and what each one measures:

Metric	What It Measures
% of controls enforced in code	CaC adoption progress
Audit preparation time reduction	Efficiency gains
Compliance drift frequency	Stability of controls
Critical audit findings	Program health check
Mean time to remediate compliance gaps	Responsiveness discipline

9.14 Final Takeaway

Compliance no longer needs to be the enemy of agility.

In SecDevOps:

- Controls are embedded as codified rules.

- Evidence is continuous, not periodic.

- Audits become routine and predictable.

- Security, engineering, and compliance collaborate seamlessly.

Compliance as Code transforms regulatory obligations into technical guardrails that protect both the business and its customers at velocity and scale.

Part IV: Practice, Patterns, and the Road Ahead

This final section ties theory to practice through hands-on labs, real-world case studies, common mistakes, and future considerations.

- Chapter 10: Practical SecDevOps Labs
  Provides hands-on exercises and real-world walkthroughs to apply SecDevOps concepts in a test environment.

- Chapter 11: Case Studies and Anti-Patterns
  Analyzes successful implementations and highlights common pitfalls that undermine security or compliance.

- Chapter 12: The Future of SecDevOps
  Explores emerging trends, AI integration, and evolving threats that will shape the future of secure software delivery.

Chapter 10 — Practical SecDevOps Labs

Theory teaches the concepts. Labs create mastery.

This chapter provides fully guided real-world labs allowing readers to apply the SecDevOps concepts, tools, and pipelines developed across the book.

Each lab will:

- Use open-source tools whenever possible.

- Be cloud-neutral or cloud-adaptable.

- Build progressive SecDevOps skills.

- Create real artifacts readers can reuse professionally.

10.1 Lab 1: GitHub Pipeline with Integrated SAST + Secrets Scanning

Objective:

Create a secure GitHub Actions pipeline that automatically:

- Scans code for vulnerabilities using Semgrep.

- Scans for secrets using Gitleaks.

- Blocks pull requests with critical issues.

Tools Used:

Tool	Purpose
GitHub Actions	CI pipeline
Semgrep	Static code analysis
Gitleaks	Secrets scanning

Lab Steps:

1. Setup GitHub Repository

- Create new private repo SecDevOps-Lab-1.

- Use a small sample Python/Node app.

2. Create .github/workflows/security.yml

yaml

CopyEdit

name: Security Pipeline

on: [pull_request]

jobs:

  security-scan:

   runs-on: ubuntu-latest

   steps:

   - uses: actions/checkout@v2

```
- name: Run Semgrep

  uses: returntocorp/semgrep-action@v1

  with:

    config: 'p/default'
- name: Install Gitleaks

  run: |

    wget
https://github.com/zricethezav/gitleaks/releases/latest/down
load/gitleaks-linux-amd64

    chmod +x gitleaks-linux-amd64

    mv gitleaks-linux-amd64 /usr/local/bin/gitleaks
- name: Run Gitleaks

  run: gitleaks detect --source . --no-git -v
```

This pipeline scans every PR automatically.

3. Test Pipeline

- *Create a PR with:*

    o *Intentional secret (AWS_SECRET_KEY).*

    o *Common vulnerable code pattern.*

- *Validate PR gets blocked.*

4. Bonus: Enforce Branch Protection

- *Enable GitHub Branch Protection:*
  - *Require all checks pass.*
  - *Require pull request review.*

Key Concepts Practiced

- Shift-left scanning
- GitHub-native enforcement
- Blocking unsafe code at PR stage

10.2 Lab 2: Secure Deployment of Containerized App

Objective:

Scan, sign, and securely deploy a Docker-based application.

Tools Used:

Tool	Purpose
Docker	Containerization
Trivy	Image scanning
Cosign (Sigstore)	Artifact signing

Tool	Purpose
Docker Hub / ECR	Image registry

Lab Steps:

1. Build Docker Image

bash

CopyEdit

docker build -t myapp:v1 .

2. Scan Image with Trivy

bash

CopyEdit

trivy image myapp:v1

- *Review detected CVEs.*

- *Rebuild image to fix CVEs as needed.*

3. Sign Image with Cosign

bash

CopyEdit

cosign generate-key-pair

cosign sign --key cosign.key docker.io/myuser/myapp:v1

4. Verify Signed Image

bash

CopyEdit

cosign verify --key cosign.pub docker.io/myuser/myapp:v1

5. Push Image

bash

CopyEdit

docker push myuser/myapp:v1

Artifact signing prevents untrusted images from being deployed downstream.

Key Concepts Practiced

- Container vulnerability scanning

- Artifact signing for supply chain protection

- Registry security hygiene

10.3 Lab 3: Terraform IaC Scan and Policy Enforcement

Objective:

Scan IaC code for misconfigurations using Checkov.

Tools Used:

Tool	Purpose
Terraform	IaC engine
Checkov	IaC scanning

Lab Steps:

1. Install Checkov

bash

CopyEdit

pip install checkov

Write Terraform Example

hcl

CopyEdit

resource "aws_s3_bucket" "bad_example" {

  bucket = "my-insecure-bucket"

  acl   = "public-read"

}

3. Run Checkov

bash

CopyEdit

checkov -d .

4. Fix Violations

- Add encryption.

- Remove public-read ACL.

- Re-run scan until clean.

5. Bonus: Automate in CI Pipeline

- **Integrate Checkov into GitHub Actions or GitLab CI.**

Key Concepts Practiced

- Policy-as-Code enforcement

- IaC secure design

- CI integration

10.4 Lab 4: Secrets Rotation and Audit Demo

Objective:

Manage and rotate secrets using AWS Secrets Manager.

Tools Used:

Tool	Purpose
AWS Secrets Manager	Secure secrets storage
Boto3 (Python SDK)	Application integration

Lab Steps:

1. Create Secret in AWS Secrets Manager

- Store database credentials.

2. Enable Rotation Policy

- **Use built-in Lambda rotation templates.**

3. Fetch Secret from Code

python

CopyEdit

import boto3

client = boto3.client('secretsmanager')

response = client.get_secret_value(SecretId='MyAppSecret')

print(response['SecretString'])

4. Review CloudTrail Logs

- **View audit trail of secret access.**

Key Concepts Practiced

- **Secure secrets storage**
- **Automated rotation**
- **Full audit logging**

10.5 Lab 5: SIEM Integration for Security Monitoring

Objective:

Stream pipeline and runtime logs into a centralized monitoring system for security correlation.

Tools Used:

Tool	Purpose
Elastic Stack (ELK)	Log aggregation
Falco	Runtime container security events
AWS CloudWatch / Azure Monitor	Cloud-native logging

Lab Steps:

1. Instrument CI/CD Logs

- Stream build logs into centralized log system.

2. Instrument Runtime Security Events

- Install Falco agent on Kubernetes cluster.

3. Build Dashboards

- **Visualize:**
  - Failed builds
  - Blocked deployments

 o Detected secrets leaks

 o Runtime container policy violations

Key Concepts Practiced

- Continuous monitoring

- Threat detection

- Compliance-ready audit trails

10.6 Metrics Dashboard Example

Metric	Description
PRs blocked by security scans	Shift-left effectiveness
Vulnerabilities remediated	Responsiveness
Secrets detected	Secrets hygiene health
CI pipeline security gates triggered	Control strength
Runtime policy violations	Production hygiene

10.7 Final Takeaway

Security isn't truly embedded until you've practiced it hands-on.

These practical labs move SecDevOps from theory into daily operational reality.

By automating, scanning, rotating, monitoring, signing, and enforcing, you build resilient, trustworthy, and scalable security systems.

"What you automate, you control. What you measure, you improve."

Chapter 11 — Case Studies and Anti-Patterns

"Most breaches happen not from a lack of tools, but from predictable human and process failures."

11.1 Why Case Studies Matter

Case studies provide practical, real-world insights into how organizations implement SecDevOps principles to address complex security and compliance challenges. They bridge the gap between theory and practice by showcasing strategies, tools, successes, and lessons learned in diverse environments. Through detailed examples, readers can better understand how abstract concepts are applied, what pitfalls to avoid, and how to tailor best practices to their own contexts. This section presents case studies that highlight the tangible impact of SecDevOps when executed effectively.

- Real-world failures provide practical lessons for SecDevOps practitioners.

- Studying failures exposes blind spots that process, culture, and automation must address.

- Anti-patterns help teams proactively recognize risk signals in their own **organizations.**

Learning from the failures of others is faster, cheaper, and less painful.

Case Study 1 — CodeCov: Supply Chain Poisoning (2021)

Incident Summary

- Attackers compromised CodeCov's Bash Uploader script.

- The malicious script exfiltrated environment variables, including CI/CD secrets and API keys.

- Over 29,000 customers potentially exposed.

Root Causes

Failure Point	Cause
Lack of artifact signing	Bash script integrity not verifiable
CI/CD secrets exposed	Tokens and credentials present in environment variables
Weak supply chain validation	Automatic updates trusted without verification

SecDevOps Lessons

- Enforce artifact signing (Cosign, Sigstore).

- Minimize CI/CD secrets in build environments.

- Verify third-party dependencies and build scripts.

- Use SBOMs to validate supply chain provenance.

Case Study 2 — SolarWinds: Build Server Compromise (2020)

Incident Summary

- Attackers injected malicious code into SolarWinds Orion builds.

- Compromised update distributed to ~18,000 customers.

- Highly sophisticated state-sponsored supply chain attack.

Root Causes

Failure Point	Cause
Build pipeline compromise	Insufficient isolation of build servers
Lack of artifact integrity checks	Poisoned builds not detected
Inadequate access controls	Lateral movement by attackers

SecDevOps Lessons

- Use isolated, hardened build environments.

- Sign and verify build artifacts.

- Monitor access to build systems.

- Use ephemeral CI/CD runners with zero trust principles.

- Implement "build of builds" reproducibility techniques.

Case Study 3 — Uber: Secrets Leakage (2016)

Incident Summary

- Uber's AWS credentials committed accidentally to a public GitHub repository.

- Hackers accessed sensitive cloud resources.

- Exposed data of 57 million users.

Root Causes

Failure Point	Cause
Secrets committed to Git	Hardcoded AWS keys
No secrets scanning in SCM	Lack of pre-commit hooks or CI secrets scanning
Long-lived credentials	No rotation policies in place

SecDevOps Lessons

- Use pre-commit secrets scanning (Gitleaks, TruffleHog).

- Manage secrets through centralized vaults.

- Enforce automatic secrets rotation.

- Disable long-lived cloud keys — use temporary credentials (STS, IAM Roles).

Case Study 4 — 3CX: Signed Malware via Supply Chain (2023)

Incident Summary

- Attackers inserted malware into 3CX software update builds.

- Malicious payload distributed via digitally signed installers.

- Leveraged developer endpoint compromise as entry point.

Root Causes

Failure Point	Cause
Developer endpoint compromise	Weak endpoint security controls
Build pipeline trust assumptions	No secondary verification of builds

Failure Point	Cause
Artifact signing gap	Signed malware trusted by downstream clients

SecDevOps Lessons

- Secure developer endpoints with EDR and MFA.

- Implement secondary attestation of build outputs.

- Use reproducible builds where possible.

- Include SBOM verification and continuous runtime monitoring.

11.2 SecDevOps Anti-Patterns

Anti-Pattern	Why It's Dangerous
"Security is someone else's job."	Lack of shared ownership across teams
Compliance after deployment	Risky late-stage fixes
Tool overload without integration	Disconnected scanners = alert fatigue
Unreviewed third-party dependencies	Supply chain vulnerabilities

Anti-Pattern	Why It's Dangerous
Secrets scattered across environments	Credential compromise risks
Long-lived privileged credentials	Persistent unauthorized access
Shared CI/CD runners	Cross-tenant attack paths
No reproducibility in builds	Undetected build system tampering

Recognizing these failure patterns early prevents major downstream risk.

11.3 Building Security Culture and Internal Advocacy

The SecDevOps Culture Loop

Behavior	Outcome
Transparency	Blameless postmortems drive process improvement
Continuous Learning	Developers embrace secure coding as craftsmanship
Security Champions	Peer educators spread practical security knowledge
Leadership Engagement	Security becomes a business enabler

Behavior	Outcome
Automation Investment	Security scales with development velocity

Culture is the most scalable form of security.

11.4 Case Study: Successful Internal SecDevOps Transformation

Scenario:

A fast-growing SaaS company struggled with escalating security tech debt and audit gaps.

Problems Observed:

- Vulnerabilities repeatedly introduced by developers.

- Security team overwhelmed with backlog.

- Audit evidence creation was entirely manual.

- Compliance slowed down releases.

Solutions Implemented:

- Launched developer-led Security Champion program.

- Embedded SAST, SCA, IaC scanning directly into pipelines.

- Implemented Compliance as Code with automated audit evidence.

- Leadership held monthly SecDevOps KPIs as business objectives.

Measurable Outcomes:

Metric	Improvement
Critical vulnerabilities	↓ 85%
Secrets leaks in PRs	↓ 95%
Audit readiness time	↓ 80%
Mean time to remediate security findings	↓ 70%

Security maturity improved without sacrificing delivery speed.

11.5 Cultural SecDevOps Maturity Indicators

Metric	Cultural Signal
% of PRs blocked by security gates	Security integration effectiveness
Security training participation	Developer engagement
Security Champion network size	Internal ownership
Postmortem transparency	Openness to learning
Automated compliance coverage	Audit readiness maturity

11.6 Final Takeaway

SecDevOps failure is rarely caused by a lack of knowledge but by a lack of discipline, ownership, and cultural alignment.

Every major breach teaches the same truth:

- Shift security left.

- Automate ruthlessly.

- Share ownership.

- Build for resilience, not just prevention.

SecDevOps isn't a project — it's a sustainable, self-reinforcing security culture.

Chapter 12 — The Future of SecDevOps

"Security evolves with technology. SecDevOps is not a destination — it's a continuous journey of adaptation."

12.1 Why We Must Always Evolve

The rapidly changing landscape of software development, cybersecurity threats, and regulatory requirements demands a mindset of continuous evolution. What is secure and compliant today may be vulnerable or obsolete tomorrow. As technologies, tools, and attack vectors advance, so too must our strategies for embedding security and compliance into the software delivery lifecycle. Embracing continuous learning, iterative improvement, and adaptive tooling is essential to sustaining resilient, secure, and compliant DevSecOps practices in the face of constant change.

Technology trends such as:

- AI-powered development,

- Hyper-automation,

- Multi-cloud, hybrid, and edge computing,

- API-driven ecosystems,

- Software supply chain globalization,

... are forcing security programs to become even more dynamic, automated, and built-in by design.

SecDevOps will not simply adapt to this future — it will be required to survive in it.

12.2 DevSecOps vs. SecDevOps — Clarifying the Terms

Definition:

Term	Description
DevSecOps	Traditional term focusing on integrating security into DevOps pipelines
SecDevOps	A deeper security-first approach where security principles guide the architecture, pipelines, processes, and culture from inception

SecDevOps flips the mindset: Security drives development velocity, not the other way around.

Key Differentiators

DevSecOps	SecDevOps
Security added into existing DevOps	Security embedded as primary design principle
May remain tool-centric	Requires deep cultural transformation
Often reactive	Intentionally proactive
Focus on prevention	Balances prevention with resilience

12.3 Future Technology Trends Impacting SecDevOps

Trend	SecDevOps Implication
AI-assisted development (e.g., Copilot)	Source code may contain ML-generated vulnerabilities — requires advanced SAST & policy checks
Hyper-automation in CI/CD	Full pipeline security orchestration must scale automatically
API-first architectures	API security testing becomes a critical DAST focus
Edge computing	Distributed security enforcement and monitoring
Multi-cloud hybridization	Unified cloud security posture management (CSPM)
Infrastructure as Everything (IaX)	Policy-as-Code will extend across networking, identity, and runtime infrastructure

Security must become data-driven, automated, and adaptive — across every domain.

12.4 AI and ML in Automated Security

*"Security tools will increasingly think for themselves —
but still require human oversight."*

Applications of AI/ML:

Use Case	Tools Emerging
Vulnerability Prioritization	ML-based false positive reduction in SAST/DAST
Threat Detection	Behavioral anomaly detection in SIEM platforms
Insider Threat Identification	Continuous user behavior analytics (UEBA)
Code Review Assistance	AI-powered static analysis, secure code suggestion
Adaptive Access Controls	Risk-based IAM and conditional access engines
Security Incident Response	AI-guided triage and playbooks

*AI augments SecDevOps — but human expertise remains
essential to contextualize risk.*

12.5 Continuous Compliance Evolution

Future compliance frameworks will move beyond periodic
certifications into real-time, evidence-driven models.

Evolution	Example
Continuous Control Validation	Automated control drift detection
Real-Time Evidence	Immutable audit logs from pipelines
Policy-as-Code Certifications	Automated mapping to ISO 27001, SOC 2, PCI
Instant Readiness Assessments	Compliance dashboards fed by CI/CD pipelines

Standards Already Adapting:

- FedRAMP Continuous Monitoring

- NIST OSCAL formats for automated control evidence

- Automated SOC 2 evidence platforms

Compliance becomes a fully integrated part of day-to-day engineering activity.

12.6 The Future SecDevOps Team Structure

Role	Function
Security Champion	Embedded in every development squad

Role	Function
SecDevOps Engineer	Builds and maintains secure pipelines
Policy Engineer	Maintains Policy-as-Code rulesets
Cloud Security Engineer	Oversees IaC and cloud posture
Identity Engineer	Manages zero trust identity architecture
Threat Modeler	Facilitates proactive risk discovery
Automation Specialist	Integrates security controls into pipelines

Security specialization will expand across multiple engineering disciplines.

12.7 Future Key Metrics for SecDevOps

Metric	Maturity Indicator
Mean time to detect (MTTD)	Threat detection responsiveness
Mean time to remediate (MTTR)	Resolution discipline
% of fully automated security controls	Automation maturity

Metric	Maturity Indicator
% of controls managed via Policy-as-Code	Governance scalability
% of code validated with AI-assisted scanning	Adoption of AI augmentation
Compliance readiness time	Audit automation effectiveness

12.8 Career Path: Becoming a SecDevOps Engineer or Architect

Required Skill Areas:

Domain	Mastery Needed
Secure Coding	Application layer vulnerabilities
CI/CD Pipelines	Build automation, orchestration platforms
Cloud Security	AWS, Azure, GCP, Kubernetes security models
Infrastructure as Code	Terraform, CloudFormation, Kubernetes YAML
Secrets Management	Vaults, encryption, credential lifecycle

Domain	Mastery Needed
Identity & IAM	Zero Trust models, conditional access
Security Automation	SAST, DAST, IaC scanning, SBOM, SCA
Policy-as-Code	OPA, Sentinel, Rego
Compliance Frameworks	SOC 2, ISO 27001, PCI-DSS, HIPAA, GDPR
Threat Modeling	STRIDE, PASTA, MITRE ATT&CK mappings

SecDevOps Engineers are cross-disciplinary leaders that sit at the convergence of engineering, security, automation, and business governance.

Certification Pathways

Certification	Value
Certified DevSecOps Professional (CDP)	End-to-end pipeline security expertise
AWS Certified Security – Specialty	Cloud-native security mastery
HashiCorp Certified: Terraform Associate	IaC specialization

Certification	Value
Certified Kubernetes Security Specialist (CKS)	Kubernetes & container security
CISSP / CCSP / CISM	Strategic security leadership

12.9 Final Takeaway

SecDevOps isn't about reaching perfection. It's about building security that scales with business velocity, adapts to change, and evolves as technology evolves.

- Secure pipelines enable safe innovation.

- Compliance becomes a natural outcome of automation.

- Teams collaborate as partners, not adversaries.

- Resilience replaces fear.

- Security becomes an invisible but omnipresent business enabler.

The organizations that thrive in the digital economy will not be those with the best security tools, but those with the best security cultures, driven by deeply embedded SecDevOps principles.

Final Thoughts

SecDevOps isn't about adding more scanners or slowing down innovation. It's about building systems, technical and human, that make secure software delivery the default.

Security doesn't mean saying no.
It means saying "Yes, but safely."

Glossary

Term	Definition
CI/CD	Continuous Integration/Continuous Delivery – a method to automate software building, testing, and deployment.
SAST	Static Application Security Testing – analyzes source code for vulnerabilities without executing it.
DAST	Dynamic Application Security Testing – tests running applications to find vulnerabilities in real-time.
SCA	Software Composition Analysis – identifies open-source dependencies and their vulnerabilities.
IaC	Infrastructure as Code – the management of infrastructure using code and automation tools.
Policy-as-Code	The practice of writing and enforcing policies in code format to automate compliance checks.
Zero Trust	A security model that assumes no implicit trust—every access request must be verified.

Term	Definition
SBOM	Software Bill of Materials – a list of all components in a software product, often used for supply chain security.
OPA (Open Policy Agent)	A policy engine that allows enforcement of policies as code across systems and pipelines.
Secrets Management	Practices and tools used to securely store and manage sensitive information like API keys and passwords.
Drift Detection	Monitoring infrastructure to detect when its current state deviates from its defined configuration.

References

1. **OWASP Foundation.** (2023). *OWASP DevSecOps Maturity Model (DSOMM)*. Retrieved from: https://owasp.org/www-project-devsecops-maturity-model/

2. **NIST.** (2020). *Security and Privacy Controls for Information Systems and Organizations (SP 800-53 Rev. 5)*. National Institute of Standards and Technology. https://csrc.nist.gov/publications/detail/sp/800-53/rev-5/final

3. **HashiCorp.** (2024). *Terraform Documentation*. Retrieved from: https://developer.hashicorp.com/terraform/docs

4. **GitHub.** (2024). *GitHub Actions Documentation*. https://docs.github.com/en/actions

5. **Cloud Native Computing Foundation (CNCF).** (2022). *Cloud Native Security Whitepaper*. https://www.cncf.io/whitepapers/

6. **The MITRE Corporation.** (2023). *ATT&CK Framework*. Retrieved from: https://attack.mitre.org/

7. **OWASP Foundation.** (2023). *OWASP Top 10: The Ten Most Critical Web Application Security Risks*. https://owasp.org/www-project-top-ten/

8. **Sonatype.** (2024). *State of the Software Supply Chain Report*. Retrieved from: https://www.sonatype.com/state-of-the-software-supply-chain

9. **Sigstore.** (2024). *Sigstore Documentation*. Retrieved from: https://docs.sigstore.dev/

10. **Kubernetes.** (2024). *Kubernetes Security Best Practices*. Retrieved from: https://kubernetes.io/docs/concepts/security/
11. **Open Policy Agent (OPA).** (2024). *Policy as Code Documentation*. Retrieved from: https://www.openpolicyagent.org/docs/latest/
12. **Microsoft Azure.** (2024). *Azure Policy Documentation*. https://learn.microsoft.com/en-us/azure/governance/policy/
13. **Google Cloud.** (2023). *Security Foundations Blueprint*. Retrieved from: https://cloud.google.com/architecture/security-foundations
14. **Amazon Web Services (AWS).** (2024). *AWS Well-Architected Framework – Security Pillar*. Retrieved from: https://docs.aws.amazon.com/wellarchitected/latest/security-pillar/
15. **Docker, Inc.** (2023). *Docker Security Overview*. Retrieved from: https://docs.docker.com/engine/security/
16. **Snyk.** (2024). *Snyk Documentation and Developer Security Resources*. Retrieved from: https://snyk.io/docs/
17. **Checkov by Bridgecrew.** (2024). *Infrastructure as Code Scanning Tool*. Retrieved from: https://www.checkov.io/
18. **Falco – CNCF Project.** (2024). *Runtime Security for Cloud-Native Workloads*. https://falco.org/

19. **Prisma Cloud by Palo Alto Networks.** (2024). *Cloud Security and Compliance Platform Documentation.* https://docs.paloaltonetworks.com/prisma/prisma-cloud
20. **ISACA.** (2022). *COBIT 2019 Framework: Governance and Management Objectives.* https://www.isaca.org/resources/cobit

Further Reading & Resources

- *The Phoenix Project* by Gene Kim, Kevin Behr, and George Spafford

- *The DevOps Handbook* by Gene Kim et al.

- *Infrastructure as Code* by Kief Morris

- OWASP DevSecOps Guidelines: https://owasp.org

- CNCF Cloud Native Security Whitepapers: https://www.cncf.io

- Official Documentation for:

  - HashiCorp Terraform

  - GitHub Actions

  - AWS Well-Architected Framework

- Online Courses:

  - *DevSecOps Essentials* (Pluralsight, Udemy)

  - *Security in DevOps (Coursera, SANS Institute)*

Appendix A: Sample Security Gate Criteria for CI/CD Pipelines

Gate	Purpose	Fail Condition Example
Code Quality Gate	Ensure coding standards are met	Code coverage < 80%, failed lint checks
SAST Gate	Block code with critical vulnerabilities	Critical vulnerability found in source code
SCA Gate	Prevent vulnerable dependencies	Known CVE severity ≥ High in any dependency
Secrets Scanning Gate	Prevent credential leaks	API key or secret detected in commit or build logs
IaC Compliance Gate	Enforce secure infrastructure provisioning	Open security group, unencrypted storage detected
License Compliance	Ensure legal compliance of open-source components	Disallowed or unknown licenses found in project

Index

About the Author

Dr. Edward K. S. Buckman, PhD, MBCI, CBCP, ITIL is a seasoned expert in cybersecurity, cyber insurance, and secure digital transformation. He holds a PhD in Technology and Innovation Management with a specialization in cybersecurity, and he is globally certified in business continuity, disaster recovery, and IT service management.

With over 15 years of professional experience, Dr. Buckman has advised organizations across finance, healthcare, and government sectors on embedding security and compliance into modern DevOps workflows. He is the founder of Nexora Press, where he leads the creation of practical, hands-on educational materials in cybersecurity, AI, and software development.

He is a frequent speaker, workshop facilitator, and contributor to industry research in SecDevOps, cloud security, and cyber resilience strategy. His vision is to help organizations build secure-by-design systems that thrive under pressure.

www.ingramcontent.com/pod-product-compliance
Lightning Source LLC
Chambersburg PA
CBHW061020220326
41597CB00016BB/1798